Cows, Cow-Houses and Milk

by G. Mayall M.R.C.V.S.

with an introduction by Jackson Chambers

This work contains material that was originally published in 1909.

This publication is within the Public Domain.

This edition is reprinted for educational purposes
and in accordance with all applicable Federal Laws.

Introduction Copyright 2018 by Jackson Chambers

Self Reliance Books

Get more historic titles on animal and stock breeding, gardening and old fashioned skills by visiting us at:

http://selfreliancebooks.blogspot.com/

Introduction

I am pleased to present another title in the "Cattle" series.

The work is in the Public Domain and is re-printed here in accordance with Federal Laws.

As with all reprinted books of this age that are intended to perfectly reproduce the original edition, considerable pains and effort had to be undertaken to correct fading and sometimes outright damage to existing proofs of this title. At times, this task is quite monumental, requiring an almost total "rebuilding" of some pages from digital proofs of multiple copies. Despite this, imperfections still sometimes exist in the final proof and may detract from the visual appearance of the text.

I hope you enjoy reading this book as much as I enjoyed making it available to readers again.

Jackson Chambers

PREFACE

This little book is written with the object of spreading knowledge with regard to cows, cow-houses, and milk, and its clean production. We ought to produce more of the fluid for consumption in this country than we do, and what we do get should be of the best. Our own sources are insufficient, and each year we import supplies from abroad which, by more effort and organization on our part, might be produced here.

Recently we took £26,566 worth of cream from France in one year. We also consumed £6,446 worth of preserved milk from Holland, Norway, and Belgium. Mr. John R. Mohler, A.M., V.M.D., one of the chiefs of the Bureau of Animal Industry in America, says: "Education is an important factor in the improvement of the milk-supply, which cannot be accomplished by laws and regulations alone. In view of these facts, it is recommended that the subject be taught in the schools; that popular articles be frequently prepared for the Press; that lectures and demonstrations be given in towns and townships; that pamphlets in plain language be prepared by the health officer for general distribution, and especially that rules and suggestions, with reasons therefor, be placed in the homes of dairymen and dairy attendants."

PREFACE

Some of the articles reproduced herein—"Improving the Old Cow-house," "Calves and their Rearing," "Milch Cattle," "Cow-houses or Byres," "Bacteria and Milk," "Variations in Milk," "Making Milk," "Drying the Cow," "Cattle Breeds and their Evolution"—have already appeared in the "Dairy Column" of *The Farmer and Stockbreeder*, and I am grateful to the editor of that publication for permission to use them again in this book. I am indebted also to Mr. John Youngs, of Chapel Field Road, Norwich, and Mr. W. H. ffiske, of Messrs. Boulton and Paul, Ltd., Norwich, for some of the notes on cow-houses. To Mr. Hugh Leader, of the St. Pancras Ironwork Co., Ltd., and others I am thankful for the illustrations.

Mr. John Burns' Milk and Dairies Bill is a praiseworthy attempt to bring about an improvement in regard to the production and distribution of milk. When its provisions have undergone criticism from practical men, a really useful and beneficial measure will doubtless be placed in the Statute Book. Cleanliness is the great thing to be kept in mind in the passage of the fluid from a healthy cow's udder to the human being's mouth. If sanitary and housing regulations be carried out by practical, broad-minded men, acting with tact and discernment, nothing but good can arise. It will be a great pity if the number of cows kept in this country be reduced owing to too stringent regulations. We want to have more of the white fluid producers kept in sound health and under good conditions. It is quite time that the cow with open tuberculosis, and visibly diseased, was as extinct as the dodo. Nothing can be said in her favour; she is a danger to her companions as well as to human beings. The Order of the Board of Agriculture

PREFACE

accompanying the Bill deals in a very fair manner with her, but most agriculturists think she should be paid for out of the Imperial funds.

The matter of cow-houses must be approached in a common-sense fashion. Many of them, especially in the country, can be inexpensively improved. The experienced veterinary surgeon should be given a free hand in the inspection of cows and byres; he is the man best fitted for the position.

G. M.

NORWICH,
September, 1909.

CONTENTS

CHAPTER	PAGE
I. CATTLE BREEDS AND THEIR EVOLUTION	1
II. BREEDS OF CATTLE KEPT FOR MILK	5
III. MILCH CATTLE—A FINE-QUALITY COW	11
IV. THE BULL	15
V. MAKING MILK—DRYING THE COW	17
VI. THE COW IN MILK—AVERAGE YIELD OF A GOOD DAIRY COW	23
VII. VARIATIONS IN MILK	25
VIII. THE AGE OF COWS	28
IX. REPRODUCTION—REMOVAL OF THE AFTER-BIRTH	31
X. CALVES AND THEIR REARING	36
XI. COW-HOUSES OR BYRES	40
XII. IMPROVING THE OLD COW-HOUSE	47
XIII. STRUCTURE OF THE COW'S UDDER	54
XIV. MILK	56
XV. MILK FOR CHILDREN	60
XVI. BACTERIA AND MILK	63
XVII. PRESERVATION OF MILK	66
XVIII. CLEAN MILK	69
XIX. ABNORMAL MILK	75
XX. DISEASES OF COWS THAT MAY AFFECT MILK	77
XXI. CHECKING THE SPREAD OF DISEASE	94
LITERATURE	100
INDEX	101

LIST OF ILLUSTRATIONS

PLATES

		FACING PAGE
I.	WILD WHITE POLLED BULL	*Frontispiece*
II.	DEXTER COW	8
III.	KERRY COW	8
IV.	DEVON COW	9
V.	DAIRY SHORTHORN BULL	9
VI.	SHORTHORN BULL	16
VII.	FULL PEDIGREE DAIRY SHORTHORN COW	17
VIII.	NON-PEDIGREE DAIRY SHORTHORN COW	17
IX.	LINCOLNSHIRE RED SHORTHORN COW AND CALF	38
X.	HEREFORD COW AND CALF	38
XI.	A GOOD UDDER	39
XII.	DEXTER COW	39
XIII.	JERSEY COW	55
XIV.	GUERNSEY COW	55
XV.	WELSH HEIFER	56
XVI.	RED POLL COW	56

IN THE TEXT

FIG.		PAGE
1.	COW-HOUSE	41
2.	CATTLE-MANGER BRICKS	42
3.	IRON COW-HOUSE FITTINGS	44
4.	IRON COW-HOUSE FITTINGS	49
5.	IRON COW-HOUSE FITTINGS	50
6.	CATTLE-MANGER	51
7.	WOOD STALL DIVISIONS, WITH IRON POSTS AND RAILS	96
8.	WOOD STALL DIVISION, WITH IRON POSTS AND RAIL	97

COWS, COW-HOUSES, AND MILK

CHAPTER I

CATTLE BREEDS AND THEIR EVOLUTION

By a slow process the chief breeds of cattle of the present day have been evolved. Their characteristics and special points have been gradually developed through past years to our own time. Capital and science have both played a big part in the progress made. Our own animals probably originated from a temperate or cold climate, as they have not the instinct of scraping away the snow to get at the herbage underneath. We know that the big cattle called aurochs, which roamed the forests of Europe in the Middle Ages, would find most of their food in the foliage and undergrowth under the trees. In more recent times Pembroke cattle on the one hand, and Welsh and Highland cattle on the other, are the types from which our breeds have arisen, and variations have been most probably introduced by crossing with imported animals of greater size. Kerry cattle have existed in Ireland from times long ago, and are considered almost typical of the oldest known native breed. Climate and surroundings influence animal characteristics. Nutritious food, under peaceful conditions and easily obtained, causes increase in size. Cold generally produces smallness in size, whereby blood circulates easily to the extremities and warmth is conserved. The skin and hair become thick, and the damper the climate, the longer the hair and thicker the skin. Wild animals in a state of nature keep truer to type than if their roaming is checked. White is generally the colour of wild bovines, and escaped cattle tend to become white. If such are kept in captivity, occasionally a black calf is born. Those tended and cared for by man arrive at maturity earlier than the wilder breeds. When roaming in flocks and herds at will females always predominate over males, the latter fighting and killing each other. The teeth appear sooner and reach full growth earlier in shorthorns than in Welsh and Highland cattle. Descent from a distinct species, climate, pasture, and food have produced in each district its special and prevalent breed. The preservation in each locality by the hand of man of those cattle he deemed most valuable has, by the process of selection, been the chief cause of the variations in British breeds. Human foresight must, however, travel along with Nature in the matter of selection. Domestic races differ from natural species. Man breeds for his own use and fancy, and this is not always for the creature's good. Thus external features sometimes prevail at the expense of constitution. When forests abounded, and before there was much arable land in these isles, and ere agriculture had advanced, many cattle were killed when the pasture was gone, and their carcasses salted for winter use. Those not killed were in most cases turned out to shift for themselves. There was no fresh meat in winter and spring. The whole meat-supply was chiefly in the hands of the rich. The monks did much to preserve the best cattle, and to give the benefit of them to their poorer brethren. Gradually, as knowledge spread and advanced, trees were cut down, the land tilled, and the marshes drained. With the advent of these things, the points sought after in the genus changed to suit the needs of man. Oxen were used for ploughing, milking qualities commenced to be considered, and animals that fattened easily, even on poor pasture, were valued.

About the middle of the eighteenth century much study and effort began to be devoted to the improvement of the bovine race; science and experience joined to work hand in hand, and the foundation of the world-wide reputation of this country for its cattle was laid. We had the natural advantages in soil, climate, pasture, and stock, and we commenced to make the most of them. At first the skill exercised in, and the credit for, the improvement of bovines rested chiefly with a few men. A spirit of emulation and striving to excel then arose, and agricultural societies were formed to improve the breeds and give prizes for competition. These spread the light, and were all for the benefit of mankind by calling and paying special attention to the attainment of those qualities needed and desired. So to-day we have breeds noted for the ease with which they put on flesh and supply excellent meat; others famous for their milking qualities; and yet again, some that produce good milk and first-rate meat. The science of milk-production as regards breed has been carried even further, and the special qualities of the milk of different kinds of cattle have been estimated and determined.

It would seem, as the years pass on, that there will be an ever-growing demand for cow's milk. The big human centres increase or keep up their population, and whilst on the one hand the physical characteristics of the female part of the nation in many cases do not improve, on the other hand the exigencies and customs of present-day life in large towns and cities lead to a slackening of the standard which great-grandmothers and grandmothers set up for the benefit of the child. It will thus be seen how necessary and desirable it is that there should be a plentiful supply of good, sweet, clean cow's milk.

CHAPTER II

BREEDS OF CATTLE KEPT FOR MILK

Shorthorns.—Durham and Yorkshire have the credit of being the counties in England from which this breed of cattle originated. In early days the animals were large in size and small in girth of the fore-quarters. They began to improve this form on the banks of the Tees a few years later, and Teeswater cattle possessed fine skin, good hair, and deep fore-quarters. Sir William St. Quintin, Mr. Milbank, and, later, Mr. Charles Colling, did much to reduce the size and improve the form of the breed. Mr. Bates and Mr. Thomas Booth in the eighteenth century assiduously cultivated the dual qualities (meat and milk) of the race, and produced cattle excellent at the pail and running into prime flesh. To-day shorthorns are the most widespread and numerous of dairy cows in this country. They have stood the test of time, and still excel as milkers, calf-bearers, and meat-producers. Perhaps if rearing was more generally carried on by selection of parents exhibiting pronounced milking qualities, a higher average could be obtained in this respect than exists at present. No breed gives such a wide field for the exercise of judgment in selecting milking-strains as this one.

The dairy shorthorn should have a long, small head,

neck thin towards the head, thickening soon noticeably, especially as it approaches the shoulder; horns wide set, lightish in colour, and curving, with age, inwards or upwards; the girth should be deep and the ribs wide. She should be well-formed across the hips, longish in the body, with thin and somewhat crooked thighs, mellow hide, and fine hair. Her udder should come well forward and teats be large. Good whole reds, pronounced roans, or clear whites, are the standard colours of typical shorthorns.

The breed is very prevalent in the North, Midlands, and London. Perhaps it does not do quite so well in the South as in the areas mentioned—a warm climate seems to be better for its meat than milk qualities.

Lincolnshire Red Shorthorns.—These are peculiar to Lincolnshire, and have recently been made distinct by selection extending over many years, and introduction of blood from the old shorthorn stock. They possess good dairy qualities, and many breeders have taken up the matter of improving their milk-yield with excellent results. They are big cattle, wide and deep in frame, hardy in constitution, and fatten easily. They should be a good red colour, and have low-pitched horns. The using of pure-bred bulls from the old parent strain on cows inheriting good shorthorn characteristics, derived chiefly from Durham in days gone by, keeps the breed up to the mark. The type is holding its own and spreading, and its qualities—viz., good constitution and aptitude to put on flesh—would seem to make it a desirable export animal where good new blood is needed and a native race wants freshening up.

Red Polls.—Inhabitants of Norfolk and Suffolk,

where they have been bred and improved, at any rate from the beginning of the eighteenth century. The original Suffolks were a dun colour, but this colour has given way to red, probably due to crossing with Norfolks. A deep-red colour is now esteemed.

It has been said that the breed originated from Galloways brought into the above-mentioned counties; but there is much evidence that they have been homed in the districts for long years.

The milking qualities of the Suffolk poll have probably received more and longer attention than in the case of her Norfolk sister, and many say that she produces more milk in proportion to her size than any other breed of cow in England.

The type is lean with a long light head, small dewlap, slender short limbs, heavy body, and narrowish loins. There should be no trace of horns in the breed.

The milk is very good for butter-making, and the animal fattens quickly and produces meat of good quality.

Jerseys and Guernseys.—Wonderfully free from tuberculosis, owing to being kept much in the open in their native homes. Noted for milk rich in fat, and acceptable as individuals in herds of milking cattle, where it is desired to keep up the fat standard in the fluid. As household cows they are prevalent, sought after, and give good milk for infants and children. They are well-proportioned and not fleshy, difficult animals to fatten, and consume large quantities of food before putting on flesh. They are consequently not economical as meat-makers. The Guernsey is a bulkier and bigger cow than the Jersey. Her colour is yellowish, or yellow and white, with a white nose and tongue. The Jersey is fawn, golden, or light grey, with black tongue, nose, and points. Both breeds are very docile, and give large quantities of milk up to within a month of calving.

Kerries and Dexter Kerries.—Natives of County Kerry in Ireland, where the breed has existed on the slopes of the Kerry Hills for long years. An exceedingly valuable cow for the man of slender means. Brought up on poor pasture and in a bleak climate, they have hardy constitutions. Their milk is excellent in quality, and by some preferred to that of the Channel Islands cattle. The Kerry is more of an animal for milk than the Dexter Kerry, which is shorter and rounder, of the dumpy type, possessing both good meat and milk qualities.

The colour most favoured is black, although red is not unusual.

They are good cattle for dealing with coarse pasture, and are said not to be at all liable to milk fever.

Ayrshires.—The dairy cows of Scotland. Externally an exact type of good milch cattle, having noticeable slender fore-quarters and wide hips. They are deep milkers with rather small teats, and horns growing upwards, outwards, and backwards. Always have some white about them, shared with red, brown, yellow, or black. A large quantity of cheese is made from the milk of Ayrshires. They are often leased out or let to dairymen called "bowers." They tend them carefully, keeping them clean, and looking keenly after their welfare. Hardy and not dainty cattle, they possess the quality of being more easily fattened than the Channel Islands breeds.

Welsh Cattle.—The black cattle of North Wales, found especially in Anglesey and Carnarvonshire, are small hardy creatures, long and low, with thighs well filled, large dewlaps, and rather heavy shoulders. Long horns are characteristic of the breed. The bull's horns are lower and wider spread than the cow's. They are hardy cattle, and excellent milkers as to quality and quantity. Colonel Platt, of Anglesey, has done much to improve the breed and to keep it in evidence. In colour the cows should be as black as possible, but they frequently have white marks on the udder. The bulls should be coal-black.

The South Wales cattle differ slightly from those of North Wales, being narrower in the back and longer in the legs, with more hair. They have finer facial lines, their horns are quite yellow, and udders black.

Dutch Cattle are occasionally met with in this country. They are black and white in colour; hardy in constitution; excellent milkers; easily fattened, with very noticeable escutcheons.

The Milking Devon is thus written about by Mr. J. H. Chick: "The milking Devon is capable of giving very good yields, never loses her aptitude to flesh, eats moderately, and gives rich milk. In all probability they will go much farther afield than they have done up to now. They are found chiefly in Dorsetshire."*

Careful breeders of **Hereford cows**, by selection of good milking strains, have in several instances produced paying dairy cattle. Better results have been obtained in this direction by cross-breeding. It is as a meat-producer, however, that the Hereford is known and excels.

By far the greater majority of milch cattle are *cross-bred*, and in their selection for milk the points of good dairy cows will require first consideration. The amount

* "*Farmer and Stockbreeder* Year Book," 1908.

of trouble required to get them for the man who buys and does not breed, and the ease with which they can be fattened and sold, will carry weight as regards profit-making. Opportunity to obtain desirable subjects and locality will therefore rule to some extent the quality of the cattle and their origins. A good cross-bred milch cow may be obtained from a likely shorthorn bull and a Welsh or Guernsey cow. A Dutch bull with a Channel Islands cow is another good cross. Milking-strain in the bull is as important to consider as in the cow, where breeding and rearing are carried on. The crossing of two pure-bred animals produces the best results. The product of later crosses is apt to be indifferent. Constitution may often be improved by judicious mating, and milk yield increased. Good coloured parents, thoroughly vigorous, healthy, sound, and possessing dairy qualities, should be used.

CHAPTER III

MILCH CATTLE

It is not always the cow that looks fleshy and plump which gives the most or best milk; indeed, perhaps a plentiful supply of the best fluid is most frequently secured from the animal of angular and lean frame. The back, as a rule, should be straight (except in Jersey and Guernsey cattle, where it often drops), the ribs well arched, giving plenty of room for the play of lungs, heart, and stomachs. The loins should be broad, the abdomen deep, and the hips wide, so as to allow adequate space for inside passengers in the shape of calves. Looked at from behind and at her own height, a good milch cow should taper from aft to fore, and her back part obscure her front region. There is no doubt that the making of milk has an effect on, and is associated with, the whole system of a cow; all her organs, therefore, should be sound, and nervous system good to stand the drain on her resources of producing calves and giving milk. A broad muzzle and fine jaws are desired. The eye should be clear, the look quiet and mild, the skin thin, the hair soft and silky. The yellow or orange tint of the skin, especially noticeable inside the ears, is considered a good indication of a rich-milk giver. The milk vein which carries blood from the udder should be large and prominent, showing that a good supply is circulating therein, and giving forth abundant nourishment to the gland. As regards milk secretion, one view is that its constituents are probably derived by transudation from the blood, and the food and water give the blood the elements transuded. Recent research, however, tends to show that the formation of milk is entirely due to activity of the gland-cells of the udder, and the chemical change brought about by the action of the blood constituents on these cells. The udder should have all four quarters curved widely, or fairly so, from the base of the teat. It should not be elongated, or have its quarters too deeply cleft, nor too pendent and flabby, denoting a laxity of the tissues and want of tone of the gland, nor yet too yielding or hard to the touch, but moderately compact and springy. If the organ is hard, the cow may have had mammitis, or "garget," and blind or obstructed teats should always be looked for, as they mean lessened milk-supply and future trouble. The teats should be large enough, a regular and equal distance from each other, with room between them to allow of being easily grasped. Supernumerary or small extra teats are said to be desirable, as indicating good milkers. The temper of the animal can often be judged when feeling her udder; and if docile and quiet—both wished-for qualities in dairy cattle—she will not object to the act.

Since Guenon made known in 1851 his method of picking out good milkers, some importance has been attached by dairy farmers to the escutcheon, which is the name given to the form taken by the fine hairs on the back of a cow's udder. The more clearly defined this form is, and the better developed by a large supply of short, silky hair, the better the cow as a rich and plentiful milk-giver. If the hair be thin and scarce, the escutcheon small and badly marked, the milk will be poor and watery even though plentiful.

Perhaps nowadays so much attention is not given to Guenon's theory as in former years, and it would be difficult to prove that some of the cows that win prizes at dairy shows always possess good escutcheons; nevertheless the rule probably prevails in the great majority of instances. The hairs of the escutcheon are not well marked on calves; but they may be discovered on close observation, and are better seen on the females than males.

An old cow should never be purchased for a milking herd, and it is not often profitable to buy one over nine years old. Troubles in her more frequently occur at calving, and the milk is poor in quality and small in quantity. A cow is generally at her best as a milker after she has had three or four calves.

Animals giving little and inferior milk, when taking into consideration the time of the year and the date of calving or duration of pregnancy, should be discarded. Where milk records are kept the sinner will soon be found, and when found she should be fattened and slaughtered.

A Fine-quality Cow

One often hears it said that a cow shows *quality*. It is doubtful whether many of those who loosely use this term know or can give an intelligent explanation of its meaning. It does not necessarily mean that she is a pedigree animal, but there is generally pure blood on one side about her. It applies to those tissues of her body that can be seen and felt, and in judging of quality these should be noticed. If she be horned, the fibres or texture of those prominences will be fine and with a fair external polish. The horn of the feet, when clean, will show a like nature; the skin will be fine, elastic near the hip, the hair silky, and the lines and ridges of the face well marked; the skin of the udder will exhibit elasticity and softness. Most of these external signs denote a good blood-supply and active nutrition. Good-quality cows in a dairy will, to a certain extent, be a gauge of an excellent milk-supply.

CHAPTER IV

THE BULL

THE bull stamps his progeny with dairy qualities if he has been wisely bred from ancestors with good points. As far as external features in his descendants go, he exercises a bigger influence than the dam. A bull out of a cow producing a large milk-supply is a far more valuable asset to the breeder of dairy cattle than one whose mother only produces a poor or indifferent yield. The best females of good form and sound constitution should be used to breed sires. There is evidence to show at the present day that the milk strain of the bull is more markedly transmitted than that of the cow. This in him will be indicated by a yellowish skin, soft to the feel, not too loose, and covered with fine hair. He should be of sound constitution, not pampered and forced to maturity too early, nor yet fat and lazy. The head should be strong and masculine-looking, the forehead broad, the eyes lively and prominent, the horns where present of fine texture and strong make. His temper should be good, and escutcheon well marked. Small teats on the purse are considered desirable as evidence of a sire likely to beget good milking stock.

Inasmuch as a bull's hocks and hind-fetlocks will perform much service under considerable strain during the stock-getting season, they should be fine and strong, and not at all puffed or swollen. He should not be flat-sided, but round and well ribbed up, with a broad chest. His belly should be well braced up by the abdominal muscles; his loins should be broad, and back strong, straight, and flat. Looked at as a whole, he should be well-proportioned, and show fine quality of bone and hair, with plenty of body space for his internal organs.

It is more important to exercise care as regards purchasing a bull than in buying a cow. If possible, a buyer should see a bull serve; he may be able to get some written assurance that his purchase is a good stock-getter, and free from all genital disorders. A bad bargain in the cow line may be got rid of without doing much harm, but an inferior bull never does a man credit; and a diseased one may cause much and widespread mischief.

It is held advisable by many astute breeders to change a bull about every three years. The knowing rearer will probably obtain better milking stock than even the well-informed buyer only of milch cows.

CHAPTER V

MAKING MILK

The food of a cow is a secondary but very important matter where a plentiful milk-supply of good quality is desired. The amount and richness of the fluid are more nearly connected with the breed and strain of the individual animal than with her food-supply. A poor and unsatisfactory milk-giver is never made a good one by any method of feeding, but the food eaten influences the quantity of the secretion more than its quality. The elements that form milk are derived from the blood and cells of the milk-gland.

The activity of the cells and the parts obtained from the blood must be kept up by suitable and sufficient food and water-supply. The latter especially should be plentiful and pure, as from 80 to 90 per cent. of water is contained in the white fluid. In all tissue change nitrogen plays an important part. This nitrogen is obtained from albuminoids or food-stuffs containing nitrogen. Other component parts of milk are obtained from carbohydrates or non-nitrogenous food-stuffs. The formation of the essential materials of milk may be influenced very favourably by right and useful feeding. The fat of milk is derived from carbohydrates (which probably furnish, chiefly through starch, one-half), albuminoids, and fat in the food. Albumin is the most important constituent of the cells of the secretory gland, and is necessary for the making of casein and milk-fat.

In all useful and satisfactory feeding for milk the proportion of albuminoids or nitrogenous material to carbohydrates or non-nitrogenous ingredients must be considered. A fair ratio for dairy cows is as 1 is to 6 or 7, and from 2 to 3 lb. per day of albuminoids in the food is good practice. Foods rich in albuminoids are oats, maize-meal, linseed, and cotton cake, bran, bean and pea meal, lucern, and, to a less degree, grass and meadow hay.

Professor Gilchrist, of the Northumberland county experimental station, gives the following rations for milch cows weighing about 10 cwt. and giving somewhere near $1\frac{5}{6}$ gallons of milk a day:

    (1) 39 lb. of swedes or 52 lb. yellow turnips.
        19 lb. oat straw.
        $4\frac{3}{4}$ lb. decorticated cotton cake.

For a cow giving about 3 gallons of fluid a day:

    (2) $46\frac{1}{2}$ lb. of swedes or 62 lb. of yellow turnips.
        19 lb. oat straw.
        $6\frac{2}{3}$ lb. decorticated cotton cake.
        $4\frac{1}{2}$ lb. undecorticated cotton cake.

With hay instead of oat-straw:

        $46\frac{1}{2}$ lb. of swedes or 62 lb. yellow turnips.
        19 lb. meadow hay.
        5 lb. decorticated cotton cake.
        $3\frac{3}{4}$ lb. Indian cotton cake.

Heavy milkers, giving about 4 gallons of milk daily,

may have an addition to their rations of 2 lb. of hay-seeds, 2 lb. of linseed cake, and 1 lb. of maize-meal, or 2½ lb. of linseed cake and 2½ lb. of maize-meal.

A feed not unfrequently given is ¾ bushel of wet brewer's grains, 18 lb. of hay, and about ½ cwt. of roots per day. This is a good one in winter: Bean-meal, 5 lb.; maize-meal, 2 lb.; bran-meal, 3 lb.; mangels, 28 lb.; hay, 21 lb. Some boil the food, mixing grains, bean-meal, and cut straw with turnips, and give oat-straw to eat. A well-known Scottish farmer (Mr. Allan, Aitkenbar, Dumbarton) gives to his Ayrshires 5 lb. of bean-meal, 4 lb. of Bibby-meal, 25 lb. of turnips, and as much straw as they want. The meals are mixed with cut hay, and fed after scalding some hours. 1½ lb. of cake to cows in full milk.

Others give chopped hay (to prevent waste), and mix it with steeped linseed, adding 3 or 4 lb. of cotton cake, or mix it with bran and boiled barley.

About 4 lb. of oats per day may be given along with bran and maize meal in half the quantities, and roots, hay, and oat-straw. Some give grass-feeding in summer, and the rest of the year hay, cabbages, and roots, and in winter cotton cake. Ensilage is now rarely given to cows. Turnips, ensilage, and cabbages are best supplied immediately after milking if the milk is not to be flavoured. When cows are on grass, about 2 lb. of cotton cake a day helps them.

To get milch cows to do their best and give most, the attendant must know each of his charges individually. He will be acquainted with her capacity for food, and temperament or habit of body.

To make the animals comfortable always means influencing the milk-supply favourably. Cows must be kept from draughts, chills, and wet. Exercise is material to their well-being, and, except in very rough weather, they should, if possible, be turned out three or four hours a day. Ceasing to chew the cud, horns and ears cold, and a dewless muzzle, indicate a cow out of order. The animal must not be worried at milking-time by hurriedly driving, or kept waiting or bellowing about when coming in to be milked. A restless cow may sometimes be quietened by letting her eat whilst being milked, and thus her yield favourably influenced. If this is done, the food should be put down some time before milking, in order to avoid causing dust in the air of the cowhouse at a time when it should be purest. All cows should be milked clean out. There are three reasons for this: secretion of milk is kept up and increased by the action of milking; the last white fluid from a cow's udder is the richest; a perfectly empty vessel is not so liable to become diseased as one in which part of the milk is left.

Drying the Cow

For the health of the cow, it is advisable and desirable that she should have a rest from milk-giving for a month or six weeks before calving. In many large herds the average length of time the cows are dry will reach as much as ten weeks. The calf in the womb benefits by this, and gains additional nutrition; the udder of the mother has time for rest and repair; she herself gains strength for the period of labour. It is not always easy to cause a gradual stoppage of secretion of milk, especially in deep givers, and some, indeed, go the time round. Others stop giving milk of their own accord, and the difficulty with them is to keep up the supply for a reasonable period. They are poor milkers, and should be got rid of in well-managed herds. All organs of the animal body need exercise to keep them in a good state, in fair working order, and so that they may perform proper service when called upon. This service may be either functional or physical. Loss of action causes parts to lose power and decline, and it is a knowledge of this law that is taken advantage of in the drying of a cow. As long as she is milked regularly and fed on milk-making food she will readily respond to the hand of the cowman. Milked less regularly, not milked out, and her supply declines or its quality depreciates. Imperfect removal of milk lessens its flow, but the practice also leaves room for the production of hard quarters or inflammation of the udder. If the gland is left entirely alone, as some may think the readiest way of causing milk-flow to cease, mammitis often arises. One sees this in a cow got up for sale with a big, heavy, overstocked bag.

The best and safest way, then, to dry her is to milk less frequently, but always strip the udder or draw away every drop of milk. Where a cow has been milked twice a day, let her be milked once. Then milk her every other day until the supply falls off.

The food necessary to lessen the flow should be dry, and not calculated to force or make the fluid. It is almost impossible to stop secretion if the animal be on grass. She will often calve well out at grass, but her udder will not get the rest it needs. A small supply of hay and straw will help to produce less milk, and if attention be paid to the state of her bowels, so that she do not come to calving constipated, nor be bound after, she will then run the less risk of milk fever.

The same rules will apply to the animal that is leaving the herd for the butcher. To dry her milk off first and then to fatten her is good procedure.

A dose of Epsom salts before commencing to lessen milk-flow drives off fluid from the system, and is right practice. Drying-drenches may follow afterwards.

CHAPTER VI

THE COW IN MILK

The food suitable and sufficient for her will be judged by the attendant, and he will know that as she advances in calf she may require more nutritious diet up to the fourth or fifth month of pregnancy, and, finally, a laxative one. Undecorticated cotton cake, when the animal is on grass, will be found very beneficial. It corrects undue laxity of the bowels during the summer months, and helps the milk-yield. Feeding should be at stated times, and if hay is given, it should be three times a day, which keeps the animal quiet and chewing the cud. The watering should always be well attended to, especially when the subject is having dryish food; but dry food and succulent food should be mixed or alternated when stall-feeding. Exercise is at all times desirable, the gentle motion in walking aiding the digestive organs, stretching the legs, and benefiting the growing inside passenger. Where cows can have access to a field, even when that in winter is only for two hours a day, they will do well. Regularity in milking is as important as proper time of feeding. Milking an hour too early often means a loss of a pint or two of milk from each cow. Inside the house cleanliness, dryness, and warmth must be observed. In winter-time water with the chill off is the best; a large quantity of cold roots is inadvisable, and these should, of course, be pulped. Warmth of the animal giving milk is most important, and she should never get wind-chilled, exhibit a staring and upstanding coat, or arched back and drooping head.

Average Yield of a Good Dairy Cow

A cow may be said to be a good milk-giver if she yields 8 to 10 quarts a day, according to breed and size. She will probably be at her best when seven years old. Herds may be said to be good where shorthorns average from 650 to 800 gallons a year; Ayrshires 550 to 600 or over (twenty-four Ayrshire cows of Mr. Allan's, Aitkenbar, Dumbarton, averaged 680 gallons); Jerseys and Welsh cattle about 500 gallons; red polls about 650; Dutch cattle about 750; and Kerries about 400. About 480 gallons a year is the total average milk-yield of cows in all the country. Many individual cows of these breeds —notably Ayrshires, Dutch, and shorthorns—may give over 1,000 gallons a year. A herd of seventy-six cows of these three breeds (mostly shorthorns) gave an average of 627 gallons for the year.*

For exhibition at dairy shows, not milking more than eleven months in the year—

Shorthorns should give about	830 gallons
Ayrshires ,, ,,	730 ,,
Jerseys ,, ,,	600 ,,
Red polls ,, ,,	700 ,,
Dutch ,, ,,	830 ,,
Kerries and Dexters should give about	440 ,,

The milk of these animals must also contain 12 per cent. of solids.

* Mr. F. C. Lloyd's Ham Farm, Beckenham, Kent (*Farm and Garden*, July 17, 1909).

CHAPTER VII

VARIATIONS IN MILK

The influences which cause milk to vary in quality and quantity are many. The amount of fat in the fluid is of primary importance to the cow-owner, and breed and strain are the chief factors in keeping up the fat standard. Two or three Jerseys in a herd of twenty or thirty cows, where milk fails in quality, will help materially. As cows advance in calf the milk gets richer in fat and less in quantity. An udder left full too long loses the fat from its contents, and the last-drawn fluid is always the richest; hence the great importance of completely emptying cows' udders at milking-time.

Heifers with their first calf are, as a rule, poor producers of lacteal fluid, both as to quality and quantity. They should have every care, and be well treated and fed in early days, so that when they are in their prime as milkers three or four years later, they will be doing all that may be desired. The supply from cows on grass will not be so rich in fat as in the autumn. Many cows calve in the spring, and this tends to make the fluid poorer for some time. Morning's standard is often slightly lower than evening's, due to the fact that cows are left longer between night and morning milking than between morning and evening. As, however, the percentage of fat in milk varies in the different cows of the herd, as well as being a matter of breed and strain, some individuals will be giving rich milk, some moderate, and some poor; on this account, the supply from the whole herd should be carefully and thoroughly mixed to keep up the average. It has been suggested that where morning's milk falls below standard, some of the first-drawn fluid from each cow should be kept back and used later on. Milk should also be stirred well when delivering. It has been the subject of some controversy as to whether food influences the quantity of fat in milk much; but there is evidence to show that where an excess of watery diet is given, or unsuitable rations, an improved food-supply will increase the quantity of fat in the fluid. For watery milk, $\frac{1}{2}$ lb. to $\frac{3}{4}$ lb. of good oats per day in the feed is said to improve fat-yield and milk-taste.

As regards the quantity of milk, long lactation periods will increase this, as well as long periods of rest before freshening and judicious feeding. Newly calved cows must be bought or brought into the herd at varying intervals to keep up the amount of the supply. The months of calving-time should vary. The business-like cow-keeper will have a sufficient supply of milk all the year round, and will want to sell the greatest quantity when it is dearest. At the time cows are taken off grass and housed, moist and substantial food should be given them. The greatest variation in the secretion takes place at the time of calving, and the contents of the udder are at that time of special benefit to the calf. About three full days after calving the milk may be used, providing the cow is clean behind, and matter from the cleansing and tail kept out of the pail. Change of pasture often has a beneficial effect on the flow, but the grass should be good and succulent, not coarse, watery, or waterlogged. In summer flies bother cows a deal, and these tormentors often seem to abound under trees, and by their unwelcome attentions lead to a loss of that repose in the animal so favourable to milk secretion. The shelter of a shed will be useful at this time. Change of weather, harsh treatment, and fright upset milk-yield, and the animals will often give down less milk to a stranger than to the regular handler.

To keep the best watch on a dairy herd and its members, fat and milk records should be kept, as well as noting the points mentioned herein. Statistics help here, as they do in most other trade matters, when intelligent reasoning and acting are based upon them. Good herds can be safely and surely built up through their aid, and the information derived therefrom. Old cows should be avoided; they seldom give satisfactory milk after nine or ten years of age.

Most of the Agricultural Colleges in the British Isles will test samples of milk for 6d. each.

CHAPTER VIII

THE AGE OF COWS

It is important that the buyer of a milch cow should be able to tell her age. He will want to know a heifer; when a cow is in her prime; how many calves she has had, and that she is not old and unprofitable as a milker. These matters can be ascertained by a study of her teeth and horns.

Two sets of teeth arise in bovines. The first set are called temporary, because only in the jaws for a time; the second lot are called permanent, because lasting, though not in their first splendour, to the finish. Those which cut the food off at the bite in the front of the lower jaw (eight in number in the full mouth, absent in the upper jaw) are called incisors, or cutting teeth. A dental pad in the upper jaw serves as a surface for the incisors to work against. The teeth grinding or chewing the food, called molars (twelve temporary, but twenty-four permanent), are situated on each side of the upper and lower jaw. The tongue of the animal, which is rough and easily protruded, pulls the food into her mouth, where it is cut off by the incisors against the dental pad, and passed on for the molars to grind by a side-to-side motion of the jaws. Temporary teeth are smaller and whiter than permanent ones. The shape of the incisors is broad, sharp at the top when new, and narrow at the necks where they enter the gums. As time passes, wear bluntens them, and both permanent and temporary teeth get farther apart as years go on.

A *calf* at birth usually has two temporary incisors, and at four weeks eight are up. At eight months the central incisors begin to wear, and at eighteen months the same thing will be noticed in all the temporary incisor teeth. At the beginning of the second year the central permanent incisors usually appear, and at about that time a heifer will be ready to reproduce her kind; the jaw itself will be widening to allow for development of the teeth. At two years and six months the second pair of incisors are up. At three years the lateral incisors are up, and the cow has usually the first ring at the base of the horn. At this time, too, she is generally brought into the milking herd, and the ring will denote that she has had one calf. At three years and six months the corner broad teeth are generally up, and are the smallest of the cutting teeth. A ring on the horn is added at each calving, and each ring represents a year, calculating three years for the first one.

With regard to the molars, or grinders, they are not easily examined without a mouth-gag; but the calf is born with one or two on each side, and soon gets six in each jaw above and below. At six to eight months the first permanent molar appears, and is the fourth in position, counting from the front. At one year and three months the fifth lasting molar is cut. At two years to two years and three months the sixth molar is up. At two years and six months the first and second lasting molars have pushed out the temporary ones and are up. At three years the third permanent molar has replaced the temporary one. At six years old all the incisors will be showing signs of wear, and will be flattened at the top with a dark, central mark. At ten the teeth will be wider apart, and the corners often bigger than the centrals. At ten a cow is old, and at sixteen to eighteen or twenty very old; but milk has been given and calves suckled at thirty-one.

Teeth are developed earlier in well-bred animals than in poor-bred ones, and bulls are usually forward in their dentition. High feeding does not, as might be supposed, hasten the development of teeth. The horns of cows for sale may be filed or rasped to get rid of the rings and disguise their age, so that it is always as well to check the signs on the horns by an examination of the teeth.

CHAPTER IX

REPRODUCTION

Heifers and bulls usually commence to be mated at about two years old. Many males begin to cover at fifteen months. They are not, at that age, in such prime or capable of producing such good results as at three years old; but it is found that they can exert their breeding-powers fairly well at that time, and profit and convenience conjoin suitably with nature. If the breeder keeps a bull of his own, he can order his cows in milk, and calves, almost at his pleasure to suit his trade requirements and supply of feed. Spring and autumn calvings are the rule, however. All the higher forms of animal life commence as cells, and it is the union of the male and female cell, and the subsequent division of their result, that produce the finished article in the shape of a calf. The male cell or spermatozoon is derived from the testicles of the bull, and the ovule or ovum from the ovaries of the cow, two of which organs in her are situated in the region of the loins. The ovule is a small body not unlike a cooked grain of sago, $\frac{1}{250}$ of an inch in size, if such can be imagined. The spermatozoon or male cell is a tadpole-like body, about $\frac{1}{500}$ of an inch long, contained in the semen or seed of the male. A head, neck, and tail may be distinguished in it under the microscope. No seminal fluid is fertile without the spermatozoa, and if bulls are used too often their seed will become unfertile.

The development in and exit of ova from the ovaries takes place by the ovules passing from the edges of the organs to their centres, and on their way getting enclosed in hollow spheres of cells, called follicles. These work their way back to the edge of the ovaries and burst, ovulation or escape of the ovules occurring. This happens at the time the cow comes on heat, or "on bulling." The ovules gain access to passages leading from the ovaries to the uterus or womb.

When the bull in the covering act sends his semen containing the spermatozoa into the cow's womb and vagina, the tadpole-like bodies begin to move by means of their whip-like tails, and go, as it were, in search of the female cell or ovum. They generally pass through the womb, and, ascending the passages leading thereto from the ovaries, the quickest spermatozoon meets the ovule coming down. Occasionally ovules get to the uterus first, and there meet with the spermatozoa. The foremost spermatozoon in the race enters the ovule with its head and loses its tail, the other spermatozoa are shut out or excluded from the ovum, and the first step in pregnancy, called impregnation, has occurred.

The fertilized ovum divides into two, four, and so on, and attaches itself to the walls of the womb. Spermatozoa do not soon die in a healthy calf-bed, and may live seven or eight days. The ova may take four or five days to pass from the ovary down the passage to the uterus. When two ovules are fertilized, twins occur; three impregnated, triplets arise.

The uterus nourishes the fertilized ovum attached to its wall, and a grouping of cells occurs, and thus skin, bowel, and nerve tissue become evident, and membranes (fœtal envelopes) form, enclosing them. When the heart of the inside passenger forms, some time later, he or she gets self blood circulation, the fœtus taking up oxygen and giving off carbonic acid gas to the maternal blood. The growth of the young in the womb takes place in a fluid or liquor, and this prevents injury to it from concussion or blows, and at the same time allows it to increase in size. This fluid, contained in one of the envelopes surrounding the fœtus, known as the amnion or water-bag, helps to dilate and lubricate the passage of the mother at the time of calving. When a cow is pregnant with young, or in calf, coming "in season" or "on bulling" usually ceases. The dam carries her passenger about forty weeks. Age, condition or habit of body, and breed are factors in the matter of length of pregnancy. Poor-bred cows generally go longer than well-bred ones, and heifers than those that have calved before. Old animals, or those in poor condition, carry their calves less time than those that are young or in good trim. Eight calendar months and a half is an early period for calving, and eleven months a late one. Twins are not uncommon, whilst triplets are. Towards the end of a normal pregnancy the calf turns in the womb, the water-bag dilates the neck of the womb and passage, appears outside, bursts, and the youngster makes an exit from the recent abode into the world.

About nine days after calving the cow comes on heat again; but it is not advisable to put her to the bull until a month or two months after the birth of her progeny. If conception does not occur, she will usually come on bulling about every three or four weeks in the summer.

When the act of mating is properly performed and

nothing results, the fault may lie with the male or female. If the bull has proved himself a sure stock-getter on other occasions, and has not been overworked, he may be excluded from consideration unless there is something obviously wrong with him, or it is suspected that he has sustained some injury at his previous service. A bull not performing his duty at two years old will have to be examined and watched. A flabby or elongated vagina and a closed entrance to the womb are faults which cause barrenness or sterility in the cow. With a heifer, it may be that her generative organs are not properly formed; she may be in too high or low a condition at the time of the covering act, or she may come of a none too prolific mother.

After calving the calf-bed slowly contracts, the afterbirth or envelopes being expelled, or they may have to be manually removed. The action of the calf sucking at the teats and nosing the mother's udder helps to bring about shrinking of the womb.

Removal of the Afterbirth or Foetal Envelopes

At the time the calf is born, the loose portion of the envelopes near at hand should be drawn gently through the exit of the womb. This precaution is wise, as, if the membranes are left entirely in the uterus, the neck of the calf-bed occasionally closes on the afterbirth and renders its removal difficult or impossible. On an average, in about four or five days the cleansing should come away. It usually stays a bit longer in winter than summer. It often sticks tightly in cows that calve before their time, and has to be carefully removed with all due antiseptic precautions if the mother's life and health are to be guarded.

REPRODUCTION

The man who wishes to do well will see to the cow getting rid of her cleansing. The prolonged holding of it by the animal means her discomfort and a great soiling of her hind parts, tail, and udder by unhealthy and sometimes putrid secretions. When the envelopes have been removed, the back portion of the cow, shape, back of the udder, and teats, tail, and thighs should be well washed down with soap and warm water in which a little antiseptic has been placed. Wipe dry afterwards, and the animal will then be comfortable, and, needless to say, the milk-supply will increase and benefit.

CHAPTER X

CALVES AND THEIR REARING

CLEAN dry bedding is best for the newly-born calf to drop on if he is produced under cover. Old, uneven floors which are not water-tight, and upon which cows may stand at calving-time, should have their surfaces levelled up or freshened once a year at least. The standing or the box should be sweet, clean, and dry. The calf's nostrils, when born, should be freed from mucus and shreds of tissue, and the mother should lick her offspring, and be encouraged to do so. The first milk of the cow, containing colostrum, is good for the little animal. It acts as a purgative, and starts the digestive tract and its glands on their healthy way. In the mother a clean udder free from dirt and vaginal or uterine discharges is desirable. Attention should be paid to the navel of the young creature, seeing that it is clean, becomes dry, and does not swell. In some cases the navel-string is tied with antiseptic tape or cord about half an inch from the opening, or the broken cord is dipped in carbolic acid or salicylic acid and collodion. If attention was always paid to the navel of newly-born calves, and to the cleanliness of the floor and cow's udder and hind parts, fewer youngsters would die of the scour. It is one of the chief missions of this book to put special accent on the matter of cleanliness. It is not desired to be faddish in this direction, but the real welfare of a newly-born calf greatly depends on absence of dirt and filth. A calf born out of doors in suitable weather stands a better chance of living than one born under cover. Not seldom the veterinary surgeon, in his visit to the calving-shed, thinks of the lines:

> "Of all the words of tongue or pen,
> How sad are those, 'It might have been'!
> But sadder these we often see,
> '*It is, but it hadn't ought to be.*'"

Fresh air, sunlight, clean surroundings and utensils, are good for young animals. In this connection the calf-bucket, among other things, must stand the test, and should be cleansed after each time of using.

Where calves are reared to come into milk as heifers, the strain of the parents should be considered, as well as what can be seen in the progeny to denote likely future benefit to the owner. The wise selection of the youngsters will have its due effect on the milk-supply.

Whole-milk during the first fortnight is beneficial, preferably sucked from the dam, and the little and often thus obtained is Nature's way with the infant calf. Gradually replace the whole-milk with skimmed milk, and then on to linseed-meal, or wheat-flour gruel, or oatmeal porridge. The gruel should always be fresh, and never the new-made put on top of and mixed with the old. In many cases, owing to the demand for and value of milk, the young must be reared by hand from birth; but in heifers it is always wise to let the calves suck for a time, as thereby the young mother is quietened and at peace at milking-time, and her udder may be the more easily milked out, without which her yield will suffer, and the milk-supply lose some of its fat.

Calves should have plenty of exercise. It is a mistake to herd or pen them up too many together, or to keep them too long in the pens.

In some cases a cow is selected or told off to rear calves, and she will often start ten or a dozen on life's way in a year. A cheap and satisfactory plan practised in not a few parts of the country is to turn two calves (her own and another) out with a heifer for about seven of the best months of the year when there is plenty of feed; then put the heifer to the bull again at the end of about eight months. The more nutritiously and liberally, within certain bounds, that young animals are fed until they become two or three years old, the better. The heifers at about the end of their second year are fit to put to the bull, and their frames and tissues should be brought up to the best pitch in these early years. The bulls generally begin to serve at fifteen months old, and at three years should be in first-rate stock-getting condition.

Cheapness in calf-rearing too closely considered is false economy, especially if the youngsters are to be kept for and brought gradually into the milking herd. The stamina and constitution are grounded and founded during the first few years of life.

As aids in feeding, cream equivalents and calf-meals have a considerable vogue in these isles, and there are many serviceable and reasonably priced articles of diet of such nature on the market. The writer has seen one of these helps suit a youngster better than another, and the attendant who knows his business will soon find this out. There is a difference in the system of calves, as in that of babes. They do not all possess the same powers of digestion, or equal ability to take up the needed nourishment from their food.

Calves are generally weaned at about five or six weeks old, and then they will often eat some hay (especially if they have been enticed gradually to do so), chaff, and pulped roots, and some cake and molasses. Crushed oats and hay are good foods.

Whatever is given in the early months must be watched in its effect. If it disagrees, a change must be made, and a close observation should be kept on the state of the bowels and growth and well-being of the body. Digestive troubles and disorders are always most frequent in early days and at weaning and teething time.

CHAPTER XI

COW-HOUSES OR BYRES

The Building.—The cow-house should be put up on dry ground if available, be suitably warm at all times of the year, easily reached by sunlight, and a sufficient air change should be possible. Dryness inside the building will be obtained by a high enough foundation of impervious material (stuff through which wet will not pass), and by walls possessing permeability to air and steam to suck up exuding vapours. Good ones may be made of hard burnt brick containing plenty of sand. The outside of walls should not be of impermeable material. Thick ones give warmth, but are expensive. Thinner walls may be made where material which stores up or retains heat is used on the inside, and substance through which warmth passes slowly on the outside. About 12 feet is a good height for a cow-house. Nine-inch brick walls are perhaps used oftenest in the erection of byres in this country. Stone and concrete walls are also erected. They may be 9 or 10 feet high for a dozen, and 14 feet high for thirty cows. Doorways should be about 4 feet wide, and have their edges rounded. Air-bricks to each stall close to the ground will help to preserve dryness in the wall and aid ventilation. Flushing the building with air (air change) helps to bring about dryness, and is important. Great heat in summer may be avoided by a projecting roof, and this will aid in protecting the outside of the wall from the weather. Climbing-plants are useful on the outside of walls, as when their leaves fall the sun will reach the walls, and the roots will draw away moisture from the supports in wet weather. Windows whose under half will open out are suitable for cow-houses. Many byres are now

Fig. 1.—Cow-House for Four Cows.

constructed with walls of strong wood-framing, morticed and tenoned together, and corrugated iron roofs, the inside of the building being lined with specially prepared felt and match-boarding. The former insures an even temperature, and renders the house warm in winter and cool in summer. Messrs. Boulton and Paul, Limited, of Norwich, erect such a building (exclusive of floor) and lined throughout at a cost of £110, and this will accommodate eight cows. One for sixteen cows would cost about £200. With the roof only lined, and the purchaser erecting the cow-house, the firm supply one as illustrated for four cows at £29, and for eight at £52. The walls are in sections for putting together by strong bolts and nuts (which are provided), for easy erection and removal.

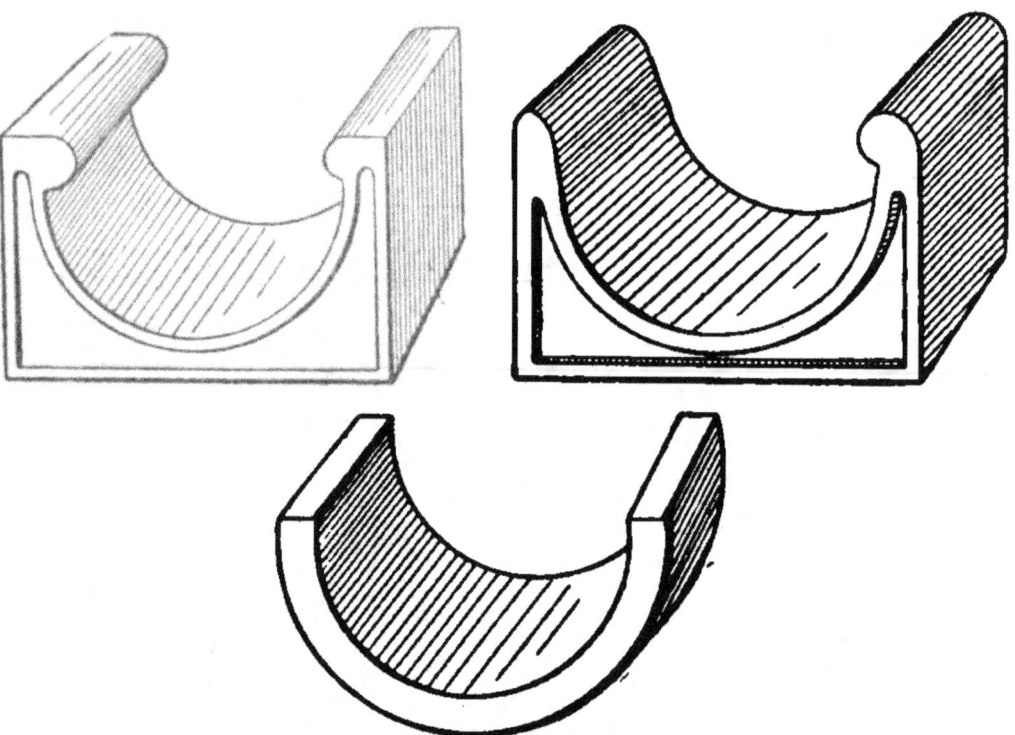

Fig. 2.—Manger Bricks, Blue Staffordshire Ware. (Hamblet and Co., West Bromwich.)

The floors of cow-houses should be of impervious material, such as well-burnt bricks grouted in cement, cement concrete, channel-bricks, flags, or large stone blocks well jointed. The stall should be level, and have a slope of about 3 inches, so that cleansing and swilling can be easily carried out. The channel or gutter should have a good fall. Enclosed roof-space should be avoided

COW-HOUSES OR BYRES

in byres; it lessens air-flow, and makes air-change difficult. The roof should be at an angle of 45 degrees. By this means air-change will be secured, even at the bottom of the walls, and the atmosphere will not remain stagnant. The same conditions as to heat-conduction should be observed in the roof as in the walls. Good roofs are made with pantiles hung on splines and bedded in laths nailed between the splines. This style of roof is called a French corrugated roof. Cement concrete for the floors of cow-houses costs from three shillings to three shillings and sixpence a square yard. Earthenware mangers are excellent, easily cleaned, and not dear. They are set in brick-work, and thus no dirt or dust collects under the manger. Butt-jointed channel-pipes, 18 inches wide, make good feeding-troughs. Iron fittings are better for cleansing than wood-work, and wood divisions fitted in an iron framework are slightly cheaper than those made entirely of iron.

Air-space should be given according to the size of the cattle and coldness of the climate. In many places cows have less than 600 cubic feet, and remain healthy. About 58° F. is a suitable inside temperature for a byre.

Ventilation may be obtained by louvre-boards for air outlets, and windows opening out at the bottom for inlets. Iron gratings on both sides of the wall, about 7 feet from the ground, are good as air inlets. A double zinc grating at the level of the mangers is a method of ventilation used by some cow-owners. There should be a ventilator for each two cows. They should be kept clean and free. Air-supply, where a tiled roof is used, may be obtained by tilting a ridge-tile here and there. In other cases air-shafts may be put through

44   COWS, COW-HOUSES, AND MILK

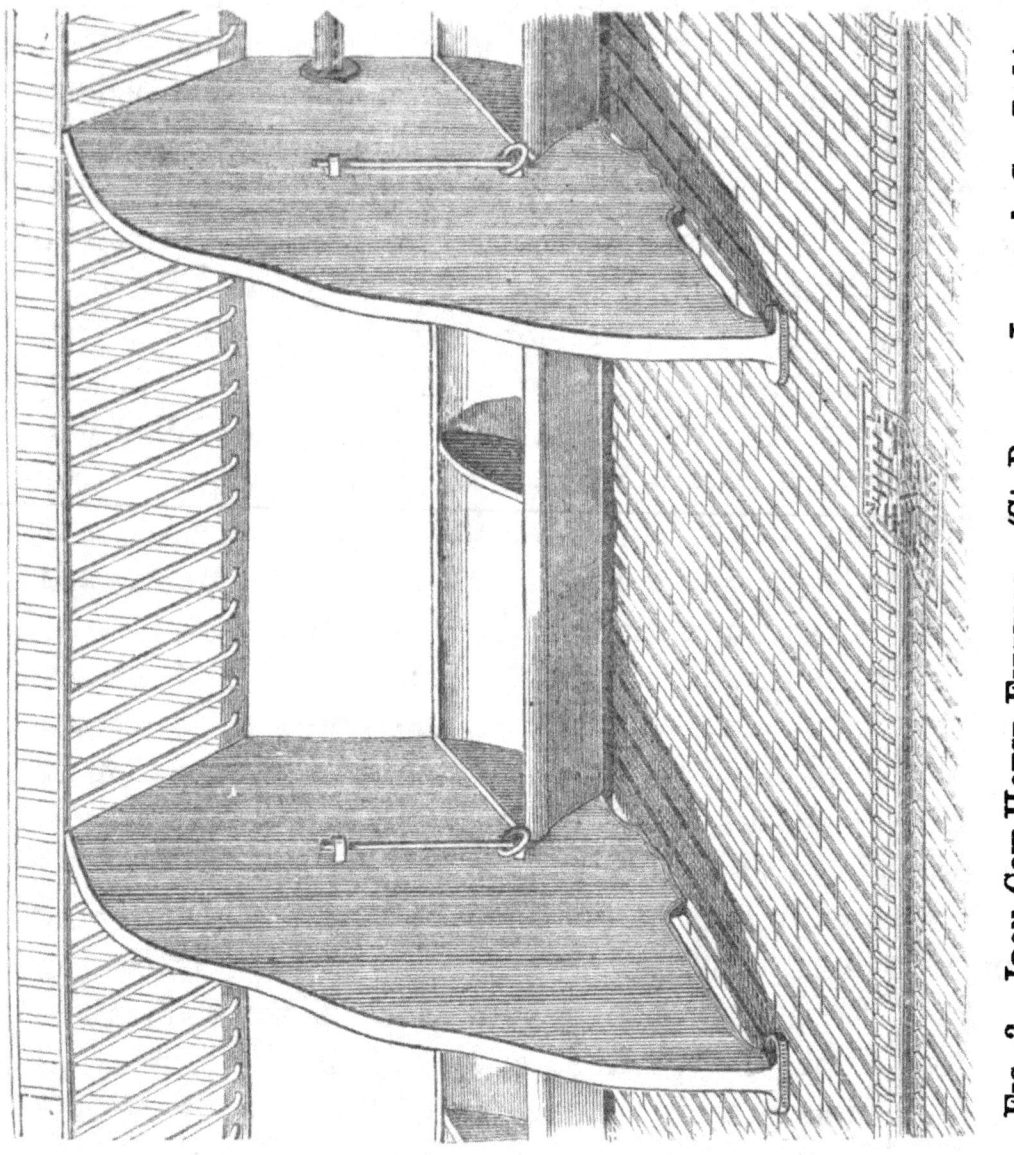

FIG. 3.—IRON COW-HOUSE FITTINGS. (St. Pancras Ironwork Co., Ltd.)

COW-HOUSES OR BYRES

the roof, and suitably covered outside. Five-inch fire-pipes as air inlets, and louvre-boards in the gables as outlets, are satisfactory.

As regards lighting, this may be from windows in the roof or wall. They may be latticed and glazed, and have sliding shutters to keep out heat and cold (Armatage*). Plate-glass windows or glass tiles may be used in the roof. Many cow-houses have no roof-windows. Glass in roofs causes heat in summer; but a thoroughly good light cannot be obtained on animals and dark and dirty corners without it. The proportion of window-space to floor-space in a byre should be about as 1 is to 15.

Stalls should be about 7 or 8 feet wide for pairs of the larger breeds of cattle. About 7 feet 6 inches from the wall, or 5 feet 6 inches to 6 feet from the bottom of the manger to the grip, is a fair length for a stall. Mangers are usually about 1 foot 3 inches high. In Holland the mangers are on the floor, about 6 to 9 inches high, and the animals chained short, so that they stretch their heads over the feeding trough. Space is thus economized behind the cows, and they keep so much the cleaner. The length of the stall in such cases will be about 5 feet 6 inches to 5 feet 10 inches.

Present regulations say that a cow should have a floor-space of about 50 feet. If there be a channel at the back of the animal, strap, chain, or fasten her so that she dungs and makes water into the channel, and is not constantly lying in her own droppings. The channel should be 3 or 4 inches deep, wide enough to admit a shovel, and with a slope away from the cows. Cleanliness will be aided by this, and the removal of dung from the cow's skin easy, because little will be

* "Cattle: Management in Health and Disease."

present. Where the animals are fastened to the side of a partition, so that they cannot get too far forward or back, removal of droppings is easily accomplished.

Double rows of stalls opposite each other, the cows standing back to back, with a long manger at each end and wide passage in the centre to allow of removal of dung and urine, are good. Grooved passages, or those of rough-surfaced cement, prevent the animals slipping.

Drains must be outside the byre, and all channels leading to them impervious. As the latter are subject to much wear and tear, they require to be well made. Hard brick is often used for them. Dung should be removed away from the cow-house at each milking-time, and dung-pits must not be too close to a byre. Regulations say they should be 60 yards away.

To calculate the cubic air-space in a building, its length, breadth, and the height of the walls must be multiplied together. For the triangular roof-space the width at the base (which is that of the width of the building), must be multiplied by half the height to the vertical angle of the roof. Add the two results together, and divide by the number of cows in the building, and the result will be a sufficiently near calculation of the cubic air-space each animal has.

CHAPTER XII

IMPROVING THE OLD COW-HOUSE

The winning of a clean milk-supply will ever be the desire of the present-day dairyman and cow-keeper. Such an end cannot, however, be attained without due regard being given to the suitable housing of the animals. Their bodily welfare, as well as the purity of the fluid obtained from them, is intimately bound up with cleansable, comfortable, and adequately roomy quarters. Cleanliness in the byre will not be hard to attain if a few modern requirements are attended to and kept in mind.

The expense connected with the improvement of many old cow-sheds need not be great, and the writer has known not a few worthy landlords who, so to speak, have set the byre in order. Many cow-houses within city and large town confines have been compulsorily altered to suit legal requirements, and nothing but good to cows and milk has resulted from the action of the civic authorities. From one aspect the work has been educative, and the movement is now spreading into the country; and what is old, out of date, and unsuitable is being replaced, made modern, and often improved beyond recognition. Many country standings contain too many cows, and not a few have too big floor-space for a cow to be kept clean. In a number of cases ventilation in the byre is deficient, and air-space of a haphazard kind; whilst floors cannot be kept clean because their structure and material render this impossible, and at the same time cause a great waste in litter. In some places floors are of chalk, rammed tight, which is not so objectionable as earth or stones of the petrified-kidney type; but impervious floors should be the rule nowadays, and are the best of all.

If there are walls with no holes in them, and windows closed tight, a byre will not be properly ventilated even if half-neck or whole doors are left open all day long. Where cows stand with their heads up to a wall there should be ventilation through the wall. This may be obtained by knocking holes in the supports and inserting perforated zinc, or iron plates, or 5-inch fire-clay pipes, one to each cow, may be used as air inlets. Windows hung on pivots are common enough as ventilators, but they should be moved intelligently, and turned to and fro as occasion demands, and not left at one angle all the year round until the iron pivots get rusty and cobwebs and dust adorn them. Where no ventilation at all exists except through the door, pivot-hung windows, or louvre-boards, may at least be put up in the gable end of the cow-house.

In order that a draught may not be prevalent, air-space must be considered along with ventilation. Eighteen feet in width should always be present in a byre, and if the cows stand with their heads up to the wall, this width should be completed behind them. Walls should be 8 feet high, or higher if width is deficient. Enclosed roof-space is objectionable, and, if it exists, should be done away with. It prevents free circulation of air, and often causes atmospheric contamination. A smooth wall-surface in front of animals

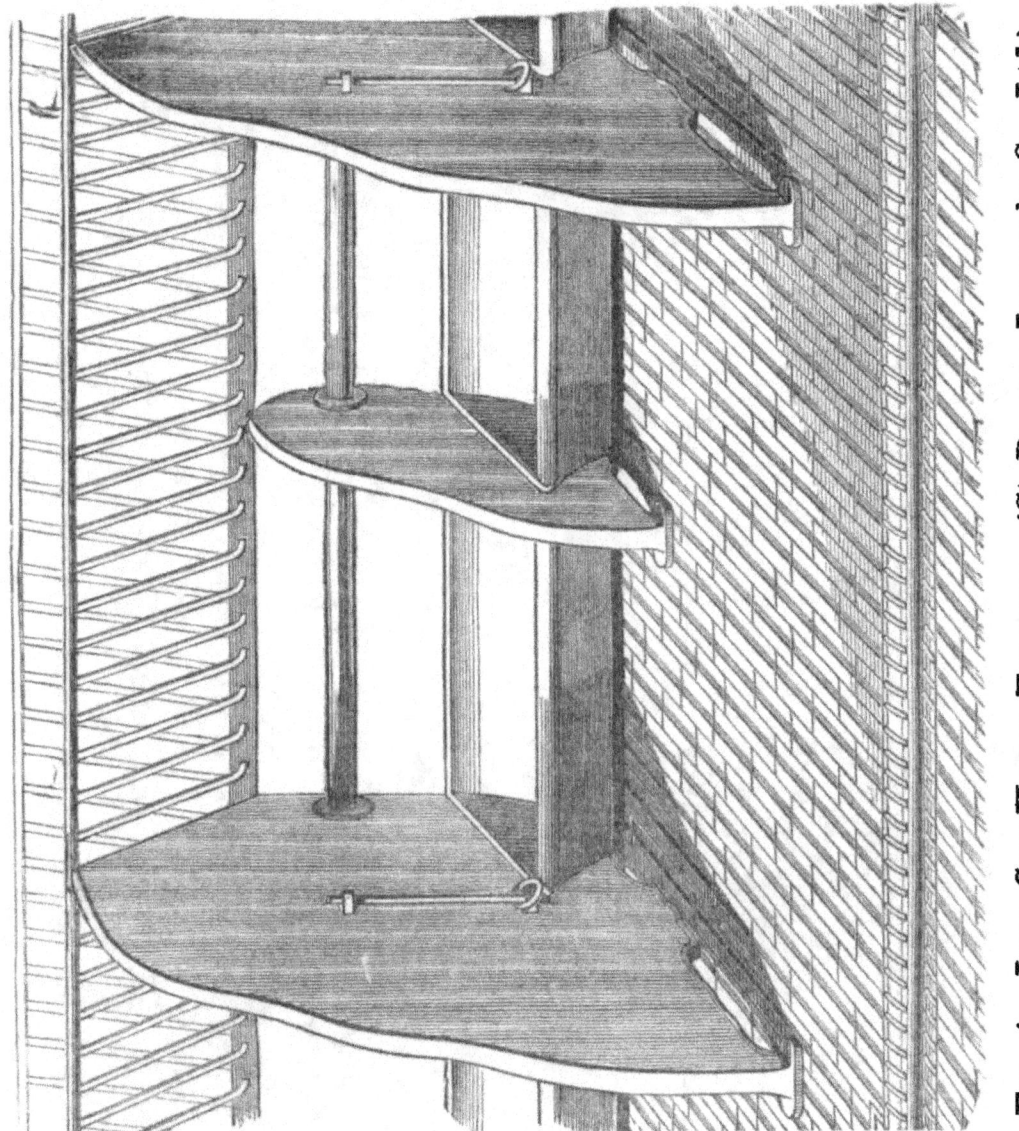

Fig. 4.—Iron Cow-House Fittings. (St. Pancras Ironworks Co., Ltd.)

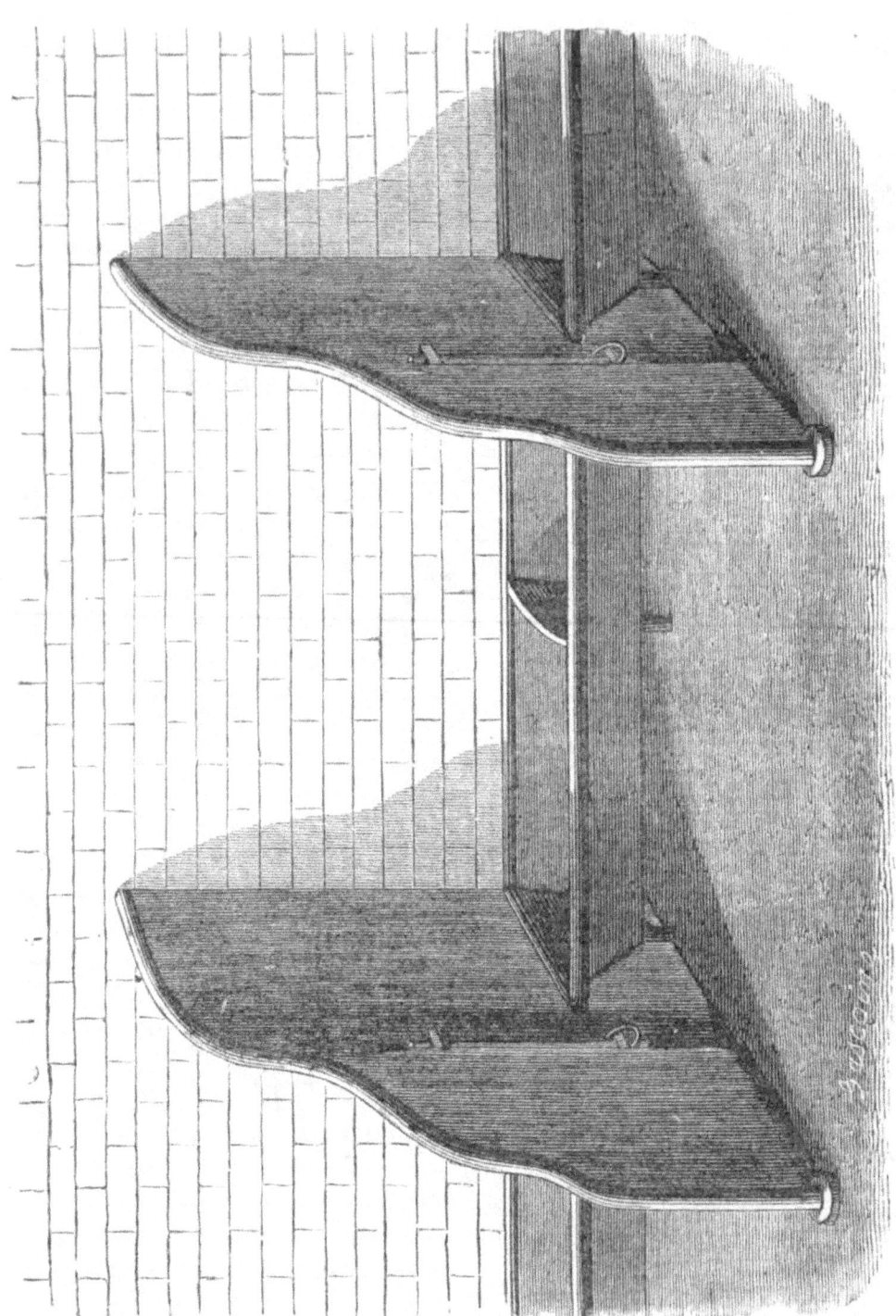

Fig. 5.—Iron Cow-House Fittings. (St. Pancras Ironworks Co., Ltd.)

IMPROVING THE OLD COW-HOUSE

standing up to mangers attached to walls is desirable. The crevices in the support may be filled up and its face smoothened, and if the structure is of brick, petrifying liquid, obtainable at 10s. 6d. per gallon, may be put on half-way up the wall or to a distance of about 3 feet, and is a great improvement. The wall can then be washed or swabbed down whenever it is fouled by the cow's breath or saliva, or bits of decaying food. Stalls should be about 3 feet 6 inches wide for each cow, and 6 feet long from the foot of the manger to the gutter. Old, unlevel floors should be made even, and nowadays impervious ones are quite the thing. Channelled bricks

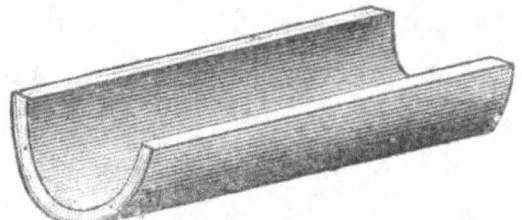

Fig. 6.—Cattle Manger. (John Knowles and Co., King's Road, St. Pancras.)

or cement concrete (at 3s. 6d. per square yard) make good floors, and even in the case of the poorest landlord and tenant a width of concrete or flag-stones well jointed can be put at the back of the cows in order to make easy the washing away of excreta and to prevent filth-besodden floors. The feeding-passages need not be impervious; well-rammed earth will do well enough for them.

In many cases, perhaps, feeding-troughs will not be altered; but if space behind the cows admits, the animals may be shifted back, and a low wall built in front of them, into which glazed earthenware troughs or butt-jointed channel-pipes, 18 inches wide, may be put. The manger is usually about 1 foot 3 inches to 1 foot 6 inches high. This existing, the cow, when lying in a 6-feet-long stall, and with a channel behind her, should be chained so that her hind parts will be kept clean. A good plan now adopted in many places is to have a manger on the floor, about 9 inches high, and chain the cows so that their heads, in a 5 feet 6 inches long stall, come over the feeding-trough. The mangers may be modelled in cement or of earthenware set in brick and plaster. Manger bricks, as illustrated, make excellent feeding-troughs.

When impervious floors are put down, channels at the back of the animals should be stipulated for, and made strong enough to stand wear and tear, and wide enough to admit a shovel. Eighteen inches wide and 4 inches deep are good channel measurements. Rotten or filthy woodwork may be replaced by new and seasoned stuff. Woodwork ought to be regularly cleansed apart from the twice-a-year whitewashing. Ironwork, being non-absorptive, is always better than woodwork, and easy to keep clean (see Figs. 3, 4, 5).

It is almost impossible to win a clean milk if cows wade through a manure-yard to the byre. The approaches should be level and dry. Holes in the wall at the ends of the channels, with a catch-pit outside, may be easily made, and the catch-pit should be cleansed and disinfected regularly. All straw- or rush-work on the inside of the roofs of cow-houses is objectionable. The nearer a smooth surface the inside of roofs, walls, and floors is, the easier cleanliness is obtained, and the better the conditions under which milch cows exist. The floor should be thoroughly cleaned and swept out after each milking, especial attention being paid to that part at the back of the cows; milk spilt or splashed about washed away; the limewashing of walls, roof, and woodwork done two or three times a year at least; and the hind parts of the cow be kept clean, as well as her udder and the milker's hands. Fowls should never be allowed in the byre.

As soon as possible milk should be removed from the cow-house to the milk-house. These things done—and they are not hard of accomplishment, nor yet very expensive—then milk will be produced under good conditions, and healthy cows will thrive. Perhaps a little bit more latitude and less rigidity may be given in country byres than in those in cities and towns, because the animals in the former case are often out in the open for many months of the year, but the improvements suggested herein will even then be beneficial, and should not be ignored by well-informed men who wish to do well.

CHAPTER XIII

STRUCTURE OF THE COW'S UDDER

The udder of the cow is suspended from the lower wall of the abdomen, and made up of external skin and connective tissue, forming holes or alveoli, like the cavities in a sponge. Around these spaces bloodvessels run and interlace. The internal surfaces of these cavities are lined with cells called epithelial cells, which enlarge considerably when milk secretion takes place, and become filled with the fluid. The milk secreted in the alveoli flows by small ducts into a little reservoir at the top of the teat called the milk sinus. There are four quarters or divisions in a cow's bag, and each compartment has its own exit by a teat and its canal. The duct or canal of the teat is lined with mucous membrane, and is controlled by a small muscle—the sphincter—which, when in working order, prevents the escape of the white fluid (see Plate XI.).

Milk Secretion.—The growth and breaking up of the epithelial cells in the alveoli is one of the most important factors in milk secretion, and there seems to be little doubt that transudation of blood and lymph constituents through the alveolar walls also enters into the matter.

Defects of the Teat Canal.—To get a good stream of milk the exit from the bag should be free and unblocked.

The defective duct may have a fault present at birth, or one which has been acquired through, or in consequence of, use or injury. The stoppage of the canal may be complete or partial. Little tumours may be present in it, or there may be a folding or wrinkling of its mucous membrane. Obstructions in the duct of the teat are frequent at the end of lactation periods. They are usually situated about ¾ inch from the end of the teat. There may be hard tissue in the canal, causing stricture and lessening its bore; vegetative growths or tenseness of the muscles, which dilate the canal as the stream of milk passes through. Closure of the canal from birth is often not so serious as in other cases, and dilation can usually be accomplished where care is exercised.

There is perhaps nothing in connection with milk production that is so annoying as blocked teats, and there is no complaint of the cow needing greater care in remedying. The difficulties experienced in setting faulty teats right can only be appreciated fully by a man skilled in the treatment of them.

CHAPTER XIV

MILK

Milk consists of water, fat, casein, lactose (sugar of milk), ash, albumin, salts, and the gases CO_2 O and N. Good milk up to standard should contain 3 per cent. of fat and 8·5 per cent. of solids other than fat. The solids consist of casein, albumin, milk-sugar, and ash. When milk is diluted with water, it becomes bluish and semi-transparent, and does not cling to a glass. Pure secretion should be white, opaque, and thick. It should consist of an emulsion of fat, a solution of albumin, milk-sugar, and ash, and salts held in suspension. The proportion of water in natural milk is about 3 in 4. It is therefore very important that a cow should have a good and plentiful supply of water; and if juicy food in good condition is supplied, milk in quantity will be formed, and a good groundwork for the quality established; 12 to 14 gallons of water a day is an average quantity for a cow in full milk to drink, and from 2 to $2\frac{1}{2}$ gallons of fluid per day the year round an average amount for her to yield. The amount of fat in milk is important; it should be about 3·70 per cent. in the best fluid. Cream is the richest part of milk, and should contain an average of 3·5 per cent. of fat. The first-drawn milk from a cow's udder is not so rich in fat as that obtained last, hence the importance of stripping cows clean when milking.

Milk secreted under great pressure, as when an animal's vessel is full, is always poorer than where moderate or low tension exists, and this explains how it is that the last-drawn milk is the richest. The secretion, when pure-won and well-kept, will retain its freshness for a long time if quickly cooled and kept cool. There should be a clean, winning, quick refrigeration, and cool preservation of the liquid.

Good milk is a perfect food for the young. When mixed with bread, it forms a good enough nourishment for adults. Alone or in an altered state (as peptonized, koumiss, or butter milk), it is almost an absolute necessity for the aged and the invalid. Milk cures for certain diseases have been practised for many years on the Continent, and with pure air and outdoor conditions, a large daily drinking of pints of milk is one of the modern ways of checking the advance of "the white man's plague," or consumption.

The fluid must not be treated lightly. It needs a lot of care to keep it good when obtained fresh and clean. It is a wonderful liquid for absorbing odours, and a splendid nourishing ground for all kinds of harmful little microscopic bodies known as bacteria. Flies love to commit suicide or have a swim in it, and when one watches some of the places the winged insects alight on before they reach the milk-bowl, it will be seen how inadvisable it is that they should wash their bodies or legs in the fluid. The household milk should be kept covered in a cold place (the coolest place in the house), and away from regions where foul or odour-tainted air may pass over it or hang around it.

Many of the harmful things in milk are destroyed by boiling. A temperature of 75° C., or 167° F., if maintained for some minutes, kills most germs in the fluid; but if the household supply is intended to be kept long, especially in hot weather, the temperature should be raised to 110° C., or 230° F., for some time. When milk is boiled, the skin on the top is due to casein, salts of lime, and lact-albumin, which are floated to the surface by boiling, and, on drying, form the skin. Boiling milk, however, causes it to lose some of its nutritive value, but not very much, and it renders it perhaps a little more indigestible, whilst the taste is insipid to some people.

If the weight of a definite volume of any liquid be compared in figures with that of an equal volume of water, the comparison is called the specific gravity. The specific gravity of milk, tested when it is at a heat of about 60° F. by a little instrument called the lactometer, should be from 1·028 to 1·035. The more fat there is in milk, the lower its specific gravity. The lactometer is a thing of glass, with a weighted bulb at the end and a graduated stem at the top. When put into milk, the depth to which it sinks is indicated by figures on the scale, and in most lactometers, by putting 1·0 before the figures shown on the scale, the specific gravity is obtained.

An instrument called a butyrometer is used to test the fat in milk. It is a sort of cylindrical bottle, which tapers off at one end to a graduated tube with a conical bulb termination. A measured quantity of milk is put into it, and shaken up with definite amounts of chemicals called sulphuric acid and amyl alcohol. The bottles are then spun round in a centrifugal machine, and the percentage of fat read off. When the lactometer is used along with a butyrometer, the percentage of solids in milk can be estimated.

MILK

The formulæ for calculating the total solids in milk and solids-not-fat are:

$$T = \frac{L + \cdot 7f}{3\cdot 8} + f;$$

or

$$T = \frac{G}{4} + \frac{6}{5}F + 1\cdot 4.$$

$$\text{Solids-not-fat} = \frac{L + \cdot 7f}{3\cdot 8}.$$

Thus, if the specific gravity of milk is 1·035, and the percentage of fat 5 (and where T = total solids, L = lactometer reading, and f = fat percentage),

$$T = \frac{35 + (\cdot 7 \times 5)}{3\cdot 8} + 5\cdot 0;$$
$$T = 10\cdot 13 + 5;$$
$$T = 15\cdot 13.$$

CHAPTER XV

MILK FOR CHILDREN

UNDER Mr. John Burns' Pure Milk Bill powers are sought to be given to sanitary authorities to set up depots to supply milk for babes. It may therefore be appropriate to write something about the fluid in connection with infants. At many places on the Continent children's milk and cows have claimed special attention. Strict veterinary inspection at short intervals of cows labelled for the supply of infants' milk is provided for. Clean and sanitary habitations, easily capable of being kept sweet and clean, are insisted on, and the winning of a pure fluid is striven after. A duty devolves on the community of furnishing a good milk for the growing race. If the secretion be won pure from healthy cows by cleanly methods and bacteria kept out of it until it reaches the babe's mouth, all that is good and necessary will have been done. It cannot be written that this will ever be accomplished when all sorts of vessels are used for the reception of milk in many and varied places, and where the fluid passes through several hands before it reaches the infant mouth.

There is much to be said in this connection for the delivery of milk in sterilized bottles as being the best method, and an effective one, at the present time. By this procedure there is some guarantee of the purity and soundness of the fluid, especially if the habitation of the cows be good and inspection regular. If the purity, cooling, collection, and delivery of milk be watched over by special individuals scientifically trained, then children's milk of sound quality will be obtainable.

There are several methods by which the ordinary milk-supply, sold without any special stipulation, is treated for infants. Perhaps the two commonest are boiling and sterilization. Boiled or sterilized cow's milk does not possess such nourishing power, nor is it so well digested, as warmed milk of good quality. The substances in the fluid that go to form nerve and bone tissue are never so well taken up by the young in boiled milk as in raw milk.

Cow's milk contains three times as much casein and more albumin and salts than woman's, and less lactose and fat. Cream, water, sugar, and lime-water are consequently often added to the food of sucklings brought up on the bottle.

Cow's milk for infants has often been dealt with to make it approximate the composition of human milk. One method is by heating centrifugalized milk to 100° C., adding distilled water, cream and sugar, peptonizing and sterilizing. Backhaus's children's milk, as made in Germany, is produced by warming skimmed milk to 40° C., adding rennet and trypsin, removing the caseous portion, heating further, adding cream and sugar, and sterilizing.

In many parts of Germany cows intended to supply milk for children are labelled as such, and periodically inspected by veterinary surgeons in order that the state of their health may be watched over. The soundness of their udders, freedom from tuberculosis, cleanliness of bodies and standings, are points which receive special attention. The milkers, too, must be men of sound health, and clean in all their methods.

On farms where milk for the young is produced, cows must be fed on certain lines. No ration which is likely to cause digestive disturbance or slight diarrhœa is allowed to be given, because changes in the nature of the milk secreted may arise thereby. The animals must be given a judicious combination of dry and succulent food, and green food in great quantity is prohibited.

A change from all dry to much green food is to be gradually brought about, and in making the change the time occupied should be about fourteen days. Full milk is always to be supplied for children. A week to ten days must elapse after calving before the milk of a recently delivered cow is used.

In Berlin the Chief of Police allows the following for "children's cows":

"1. Meadow hay, well got, of good colour and aromatic smell, and containing few objectionable herbs; not mouldy, damp, dusty, or containing fungi.

"2. Straw from cereals; it must not have a fusty smell, nor contain fungi or injurious herbs.

"3. Good unadulterated, unspoiled rye, and wheat bran.

"4. Good, sound, sweet, bruised oats, barley and rye.

"5. Linseed meal of good quality.

"6. Dry brewers' grains, above suspicion.

"All other foods are forbidden."*

About 400,000 children have hitherto died in Germany in a birth figure of 2,000,000. Milk depots for the sale of the fluid for children have been established in Berlin, Darmstadt, Stuttgart, Munich, Wiesbaden, and the kingdom of Prussia, and the heavy mortality rate is decreasing.

* Rievel, "Handbuch der Milchkunde."

CHAPTER XVI

BACTERIA AND MILK

BACTERIA are the smallest and most minute forms of plant life. They exist practically everywhere, except very deep down in the earth, at the top of high hills and mountains, and far out on the ocean. When they enter the bodies of men or animals many of them produce disease, and the fight against them has to-day become the most important one that sanitary science is engaged in. When these little organisms attack matter from a creature source outside the body they produce putrefaction, fermentation, and chemical change. Like all plants, they require food and moisture for their growth and multiplication, and although sunshine kills many of them after a certain time, yet warmth causes them to grow and multiply rapidly, whilst great cold kills them, and coldness hinders their increase and spread. Advantage is taken of this fact to check bacterial growth in milk by refrigeration, which is generally practised with all fresh-drawn milk nowadays, by which process it is cooled to about 40° C.

Germ life has many forms: some of the organisms are round in shape and occur singly, others live in pairs or exist in chains, bunches, or have whip-like tails, with which they move or swim. Rod-shaped and spiral bodies, known as bacilli, are also common.

Bacteria grow in different ways. They may develop in length, or throw out bodies called spores, which grow into adult organisms. The young offshoots are not so easily killed by great heat or cold as their parents. Micro-organisms multiply quickly, so that a small number of them under favourable circumstances and in a few hours may produce progeny almost as numerous as the grains of sand on a big seashore.

Milk from a healthy cow with a sound udder is free from germs, and in proportion as it is drawn in a pure air, protected from dust and dirt, cooled quickly, and kept at a cold temperature, the longer will it remain wholesome and good. In the end germs almost invariably gain access even to pure milk, and it is a splendid seed-bed for them. The action of bacteria on the fluid is not, however, always harmful. Some of them seem to neither hurt nor improve milk, but remain, as it were, neutral. A few are useful, and turn the fluid sour, and enable butter and cheese to be made. Unfortunately, however, the souring of milk enables many harmful organisms to make a home in it, and to grow and develop largely.

Many germs that may reach fresh-drawn milk in a cow-house cause direct injury to the fluid, and yet with care and intelligence numbers of them can be excluded. Pure air, cleanly surroundings, litter, animals and attendants, and suitable and clean utensils, greatly reduce the enemies of milk in number, and make room for the furnishing of a clean supply, of which any dairyman or cow-keeper may be proud, and with which any human being may be greatly satisfied and benefited.

In considering the impurities that may get into milk in a byre, one should remember the natural conditions, and some unnatural states that may occur therein. Wherever cow's milk is produced, there is always the excreta (dung and urine) of the cows to deal with in the same building; indeed, it is this nearness of dung and urine to milk that is perhaps the prime matter for consideration with regard to maintaining the purity of the fluid.

To keep manure out of milk, the cleanliness of the cow's tail, udder, and flank will need looking to. When the cows are casting their coats, measures should be taken to keep the hairs out of milk, and scurf, or what is technically known as epithelial cells, must be excluded. The soil of the floor, where that is not impervious, the water-supply, proximity of sewage, and soundness of food, will all need supervision. The wide-awake man will know where the enemies of milk purity lie and arise, and adopt all rational measures to keep them from the milk-supply.

When a surgeon is going to perform an operation, he has his instruments cleaned and boiled; he washes his hands in an antiseptic, and puts on a clean overall, previously rendered germ-free by heat. His patient is washed, shaved if necessary, and put on a well-cleansed table. The air of the operating-room is kept as pure as possible. What is all this done for? Simply to prevent contamination of the surgical wound by organisms—in other words, to prevent infection by bacteria or bacilli. A scientific attempt at rigid cleanliness for the benefit of the patient is being made. Just as the surgeon approaches a patient, so in a less particular and minute degree should milk be approached, the milk being the patient, and the milker the surgeon.

CHAPTER XVII

PRESERVATION OF MILK

If milk is subjected as soon as possible after it is drawn to a temperature of about 40° F., the growth and increase of any organisms in it receive a check. The processes by which milk-warmth is reduced are cooling and refrigeration. A receptacle above some horizontal tubes receives the fluid as it is drawn from the cow, and the liquid passes over the tubes, and is received into a can at the bottom. Cold water circulates in the tubes, and as the stream of milk flows over their surfaces, it is cooled and exposed to the air, or aerated, by which odours are expelled. The colder the water is, and the slower it circulates through the cooler, the better for the milk. This is the general way of reducing the temperature of the secretion, and it is simple and fairly effective.

In refrigeration proper the cold produced is greater than by the cooler, and the lacteal fluid may be reduced to a temperature of 38° F. thus. This method consists of a freezing mixture in brine, which becomes vaporized; the vapour is sucked up, compressed, and condensed, and the milk run over cold brine in a cooler.

Cooling and refrigeration hinder the action of bacteria.

A tester for unclean milk is a recent invention by Dr. Gerber, and as exclusion of dirt in which organisms abound is a great factor in milk preservation, it may be mentioned here. A glass vessel is used for this with wire gauze at the bottom, over which a cotton-wool disc is placed. The milk filters through the cotton-wool, and the impurities are left on it, and from an examination of the disc the proportion of dirt may be ascertained. Farmers and dairymen by this means may ascertain whether proper care is being taken in the production of the milk-supply.

In *pasteurization* the milk is heated by steam to a temperature of 148° to 170° F., then cooled down to 50° F.

In *sterilization* the fluid in bottles is subjected to a moist heat of 118° F. in an enclosed iron vessel for half an hour, and is then gradually cooled and the bottles corked with sterilized stoppers.

Buddeization is another method now used to preserve milk. The fluid is heated to 122° F., put through a centrifugal machine, and peroxide of hydrogen is then added. This treatment of milk originates from Denmark, and is said to kill both bacilli and their spores.

Certain chemicals are occasionally added to milk to preserve it. Perhaps the commonest are boric acid, salicylic acid, and formalin. These may be detected roughly by simple tests.

Boracic Acid.—Evaporate two tablespoonfuls of milk to dryness, and add a few drops of strong sulphuric acid to the ash. Put on some spirit and light, and if boric acid is present a green flame will be produced.

Salicylic Acid.—A few drops of ferric-chloride solution added to milk containing the acid will produce a violet colour.

Formalin.—Put some milk in a glass tube, and pour a

few drops of strong sulphuric acid down the side, and a violet ring forms round the edge of the milk if formalin is present. All chemicals in milk make its consumption harmful, especially to children.

Condensed milk, useful to travellers and armies in the field, is produced by heating milk in hot water until it reaches a temperature of 150° to 175° F. Sugar is then added, and it is subjected to the action of steam for about three hours. This drives off a lot of its water. It is then cooled, put into small cans, sealed, and sold in the form known so well.

CHAPTER XVIII

CLEAN MILK

To insure clean milk the cow, water-supply, attendant or cowman, utensils, litter, milk-house and wash-house will come under notice.

The Cow

(a) **Grooming.** — A curry-comb and dandy-brush should be kept for this. Where cows are kept clean, an average of two minutes per day with curry-comb and brush will keep them so. If the dung is caked on the cow, and scraping or brushing will not remove it, it ought to be shaved off. Dung and milk are both plentifully produced by cows, and every cow-keeper and cowman should try to do everything possible to prevent the one getting into the other. The animal's flank and buttocks should be brushed, all loose hairs and dirt removed, and washed and sponged if soiled. The udder should be cleaned, or, if dirty, washed with soap and water (creolin or carbolic soap is good both for udder and milker's hands), or sponged and dried. If teats are left wet they may chap. A clean tail, prevented from flicking dung about during milking, is desirable. A tail with dung caked on it may be washed with soap, soda, and hot water, then brushed out and dried. In some cases the tails of cows are fastened to a hind-leg whilst milking proceeds.

(b) **Damp and Dry Milking.**—Whichever method is practised, the udder and flank should be clean to start with, and the hands in wet milking not moistened with milk. Some men hold that the first milk from every teat should not enter the pail. Organisms gain access to the canal of the teat between the milkings, and these tend to make milk decompose. In past years it was often the custom to milk the first stream of the fluid on the floor with the object of washing the duct of the teat out, or, as some say, to propitiate the gods. The custom had reason in it, and if the abstracted milk had not been milked on the floor, the act would have shown the wisdom of our ancestors. If there be any loose hair, dust or dirt on the cow, it may be caught in the pail when this is put too far under the cow in milking. It is said that cows can be milked quicker with moist than dry hands, and, given clean hands, and suitable soap and water used to wash them and moisten them, damp milking is then performed under the best conditions, and is quite satisfactory. Dipping hands into the milk and milking into the hands are both bad customs. All milking must be gone about quietly. The cow-house should be thoroughly cleaned and aired before milking begins, and the floor sprinkled with water or lime-wash in hot weather.

(c) **Diseases or Disorders of the Cow in which Milk should not be Used.**—1. Where blood or pus is discharged from any quarter of the udder—*e.g.*, actinomycosis and botryomycosis, or in recent injuries of the vessel.

2. In mammitis or inflammation of the udder.

3. Where the cow is a waster or piner. An indurated udder ought to be regarded as suspicious, and ought to be examined by a veterinary surgeon. Tuberculosis and

its danger will, however, never be avoided until farmers and cow-keepers are fairly compensated and the diseased animals slaughtered. A tuberculous cow should not be left with her mates, but put by herself.

4. In diarrhœa.

5. In any disease of the cow causing rise of temperature (fever).

6. In septicæmia (blood-poisoning) and pyæmia (pus or matter absorption).

7. Until three days after calving, or some say until the cow has cleansed.

8. In milk fever.

9. In foot-and-mouth disease, anthrax, rabies.

10. In cow-pox or vaccinia, which may be conveyed from cow to cow by the milker's hands, and also inoculate or vaccinate him.

11. In red-water.

There is no disease of the cow that occasions scarlatina in the human subject. Cholera, typhoid, diphtheria, scarlatina, erysipelas, and syphilis may be spread by milk if it comes into contact with the germs or organisms of these diseases.

Water-Supply

There should be a tap and basin in every byre for the milker to wash his hands, or a bucket of clean warm water, towel, and suitable soap. Water for cows should be plentiful and of good quality, clean, and preferably not from a pond; soft rather than hard, and in winter may be given with the chill off. Where cattle go down to drink at a moving stream uncontaminated by sewage, the water-supply may be satisfactory. They often keep well even when drinking filthy liquid out of a pond, but adherent filth from the pool may then get into the milk-pail from their udders or the under surface of their bodies. Where the water-supply is bad milk often goes sour quickly. Ponds, if used, are better situated in the sunlight than the shade, and sewage should be prevented entering them. Pits require their bottoms well dredged every three years, and to be cleansed when stagnant and slimy. Good wells and springs furnish better drink than rain-water, but the last is a source of supply often used when collected from roofs and stored in tanks. Aim at having a plentiful supply of the best available water near to a cow-house or conducted right into it. In country byres an oil-drum with tap attached may be used for the milker to wash his hands, and the dirty water allowed to run away as he washes.

The Attendant or Cowman

A thoroughly healthy milker is best, with head, cap, clothes, and hands clean. The hands may be washed with an antiseptic soap and hot water before starting to milk, and it is a good plan to wash them after milking each two cows, or some say after each cow. Clean overalls for milking, as worn in some parts of the country, notably Dorset and Wilts, are desirable things.

Utensils

Vessels that have contained milk should be washed in warm (not hot) water as soon as they are empty, refrigerators especially kept clean with streaming warm water as soon as finished with. They are handy close to the cows, but pure air should surround them. Milk cans or churns may be rinsed with warm, then cleaned with a brush and hot water, to which soda has been added (1 lb. to 20 gallons), and finally subjected to boiling water or steam. Steam, where available, is unequalled as a purifier; no bacteria will live in it, and it penetrates chinks and crevices. Once a week scour all utensils with fine sand; discard rusty cans. Carefully cleanse strainers and sieves by brushing and steaming. Where a boiler is used, things should not be washed in it, but clean water only found therein. Bass brushes of different shapes and lengths used for scrubbing cans and pails need scrupulous cleanliness. All milk utensils should be aired well in a dry atmosphere, free from breezes wafted over a manure-yard, away from fowls and pigs, and cans put on a lathed table to drain, and sunned if possible. The using of cloths to dry milk vessels is objectionable because the cloths are often unclean, but if they cannot be dispensed with, as frequently occurs in town dairies, a rinsing in warm water, washing out in borax solution, and boiling will make them satisfactory.

The chief object of having milk-cans with narrow necks, and narrower at the top than bottom, is to cause as little surface of the milk as possible to be exposed to air, dust, and infection.

Litter

This may be of straw, cut straw, sawdust, or chaff. Soiled litter should be removed twice daily before milking, and damp bedding dried, and fresh clean stuff often supplied. Where cows are tied so that they dung and urinate into the channel behind them, labour in keeping the litter clean and dry is saved, and the air of the cowhouse will be so much the purer.

The Milk-House

When milk is stored, it should be in a building cool, dry, clean, and well ventilated, and with a floor of hard tiles well jointed, or cement concrete. The ceiling and walls should have air-space, and pigs and fowls be kept away from the milk-house.

The Wash-House

Best with an impervious floor. The boiler or copper should be kept clean, no water left in it overnight, and kept covered as much as possible, whether full or empty; it should be used solely for clean water, and not to wash milk utensils or milking-pails in.

Racks to allow cans to drain will be needed in the wash-house, but the airing of the vessels should be done elsewhere. The wash-house and its contents require to be left as clean and dry as possible after cleansing operations, and particularly so overnight.

CHAPTER XIX

ABNORMAL MILK

Abnormalities in Milk may be due to bacteria, unsuitable food, or disease of the cow.

The following are the most common defects in the fluid:

Slimy, Stringy, or *Ropy Milk* is due to catarrh of the udder in the cow, unsuitable brewers' grains, and occasionally or secondarily to a faulty strainer. Butterwort, when eaten by the animal, is said to cause it. This plant, the botanical name of which is *Pinguicula vulgaris*, grows on mountain-sides and damp, stony ground, and has a large bluish-purple flower and spreading ovate leaves. Milk kept in a damp milk-room or near stagnant pools will go stringy. If milk is slimy, each cow should be tried until the culprit is found.

Blue Milk is due to the blue milk bacillus. It is said also to be brought about by impure drinking-water, diseased grasses, or air tainted with putrefying animal matter. Cows eating large quantities of flowering rush (*Butomus umbellatus*) may give blue milk.

Bad-smelling Milk is caused by indigestion in the cow, or where she has eaten foul turnip-tops or badly saved clover. It may also be due to dirty utensils.

Red Milk occurs through blood finding its way into the fluid from a leaking vessel in the udder. It is sometimes seen when a cow has red-water; it may be due to bacteria. If due to organisms, the milk, if left to stand, will go red on top; if due to blood, redness will be seen at the bottom.

Bitter Milk may be due to bacteria or food. If due to food, such as lupines or wormwood, the taint will be noticed immediately after milk is drawn; if due to bacteria, such as those of the peptonizing class, the bitterness develops on standing.

Watery Milk is occasioned by watery and frozen food or too wet pastures.

Bad-tasting Milk may be due to cows cleansing badly after calving.

Inasmuch as many of these abnormalities in the secretion may be communicated from one cow to the whole milk-supply, it is important to find the offender that is causing the trouble, and this can often only be done by testing the milk of each animal separately. Cows giving faulty milk should be attended to last in the herd, and the pails used for them should always be washed immediately afterwards.

CHAPTER XX

DISEASES OF COWS THAT MAY AFFECT MILK

APART from udder complaints, many diseases of the cow cause changes in the colour, consistency, and taste of milk. In most affections supply is seriously lessened, and in some cases the flow ceases entirely. The composition of the fluid will be injuriously affected through the gland cells of the udder suffering from noxious products formed in high fever. Poisonous matter produced in the intestines in digestive troubles is absorbed and excreted by the udder, causing injury to the cell tissue. In red-water, tuberculosis, foot-and-mouth disease, etc., the colour and consistency of the secretion are changed. The percentage of fat in milk may vary from day to day in foot-and-mouth disease, and the taste become disagreeable and salty; in lung troubles the amount may rise to three or four times its natural quantity, or it may decline considerably in milk fever. If there be too much albumin in milk, which occurs in some diseases, the fluid will coagulate quickly on boiling (Rievel).

Tuberculosis

Tuberculosis in cattle is caused by a bacillus or minute organism only to be seen under the microscope. This small body acts on a gland, tissue, or organ, and produces a little knot or nodule, called at the beginning a tubercle. The organism in bovines is much like that in human beings, but it has special features which distinguish it, and which have caused it to be called by scientists the tubercle bacillus of bovine type. The knots or lumps which this germ causes vary in size from a small shot to an apple. After forming, they may caseate or become cheesy, form an abscess or cavity filled with pus or matter, or turn into ulcers. The ways that cattle take in the disease are by means of the mouth or through the nostrils. By far the most common mode of infection is through the mouth and digestive tract. A diseased animal coughs on the food or drinking-water of its neighbour, and the latter swallows the infective material; or a subject, when licking the wall, manger, or body, takes in the organism which an affected beast has thrown out. The surroundings of a "piner" or "waster" where the atmosphere is moist may be great sources of infection. There often arises more danger from them than by the taking in through the breath of dried bronchial or nasal discharge.

A common situation of the tubercles is on the thin membrane, covering the lungs and lining the chests of cattle, called the pleura. The bacilli get into the bronchial tubes, and thence to the windpipe, and are discharged by the mouth or nose. Some of them are swallowed and pass out in the dung, and other animals contract the disease from the droppings, bits of which may be thrown by the tail about the cow-house or splash up around. The kidneys and liver are not unfrequently the seat of disease; the calf-bed, vagina, and udder also suffer. These organs pass out the organism, and the risk of one animal taking the disease

DISEASES OF COWS THAT MAY AFFECT MILK

from another through these sources is a very real one. A tuberculous udder always discharges tubercle bacilli in the milk, and a cow with open tuberculosis generally infects the secretion. The soiled tail of a diseased cow may especially help to scatter infection.

It is estimated that from 20 to 30 per cent. of cattle suffer in a more or less degree from tuberculous trouble. To eradicate the disease all at once and produce ideal herds entirely free from the scourge must therefore be a slow process, if it is ever attainable by practicable measures. In many cases the affection, which is generally of slow growth, is at first confined to the animal body, and few, if any, bacilli pass out of the system to spread the mischief. When the complaint advances, the organisms are thrown out from all the openings of the subject — the mouth, nose, rectum, vagina, and the teat canals of the udder.

It is important that the cow-owner should be able to recognize the advance of the disease. For the sake of the other members of the cow-house, no less than from the point of view of infection of human beings through milk, when he does recognize it, he should separate and isolate the affected animal, and cease to use her milk. The signs of the progress of the complaint in cattle are shown by a frequent cough, especially if the animal rushes out into, or is exerted in, the cold air. The cough, too, may be heard when she is drinking cold water. Food may be taken well at first; then coughing becomes more frequent and painful, and the throat may swell. Colicky pains may be exhibited, not unseldom there is diarrhœa off and on, and slime, pus, or blood may be present in the dung. If the calf-bed or vagina be the seat of disease there will be a dirty, slimy, yellowish discharge from the shape, whilst an excessive flow of urine may point to kidney trouble, or the water may be thick and smell very strongly.

The way that milk generally gets infected in the cow-house is through the udder, or from the dust and manure in the shippon. During the process of milking infective matter may get into the pail from the cow's tail, flank, and soiled udder. *The signs of tuberculosis of the udder* will generally be shown by a painless and not very tense thickening of one or both hind-quarters of the vessel at the top. This thickening slowly becomes hard and knotty, and may be well defined. It causes the remaining portion of the quarter not attacked to decrease in size. In other cases, if the bag be milked out, little hard knots may be felt in it. The glands high up at the top of the udder, felt from behind, will be found hard and enlarged.

The milk itself is often thin, watery, of yellowish hue, whilst later on little white flakes and lumps may be found in it. The general state of the cow, when the disease advances, will easily be shown to be bad. She will fall off in condition, spaces that should be round and fairly well filled out (such as those of the neck, between the ribs, and flanks) sink in, and finally she will be reduced to a veritable bag of bones. If the udder be diseased, the milk will certainly be affected; and if the fluid from a bad bag is mixed with good, sound milk it will impart its harmful qualities to the lot. From what has been written, one thing may be gathered, and that is that the animal showing visible signs of the disease, and with open lesions of the lungs, throat, bowels, kidneys, uterus, or udder, will be discharging the makers of the disease, and will be a very powerful

DISEASES OF COWS THAT MAY AFFECT MILK

spreader of it. She should be slaughtered, and her milk certainly discarded. The writer thinks, in the interest of the farmer and the cow-owner, no less than in that of the public, some universally agreed-on means should be found to get rid of the chief offenders from the milking herds of our land. When this is done, the first real, firm step forward to oust the bugbear will have been made. The disappearance of the chief spreaders of disease will clear the way to a gradual and practicable means of dealing with other subjects less dangerously, though widely, affected.

The tubercle bacillus of bovine type has certainly been found many times in cases of consumption in adult human beings and children. The writer has at present before him records of the organism being found in twenty-five cases out of ninety-nine investigations made on human beings suffering from consumption. From the human point of view, therefore, no less than from that of the cow-owner, some scientific methods should be commenced to fight the disease. It is estimated that 2·5 per 1,000 individuals die of consumption in England and Wales in a year. Close crowding, confinement, and bad air make the disease progress; fresh air, exercise, and sunlight help to keep it away.

Fat, casein, and lactose all diminish in the milk from a tuberculous udder. The production of the fluid is by filtration and secretion, and the cells of the lining membrane of the vessel are intimately concerned in both these processes. They suffer considerably in this complaint, and, when attacked, lose their integrity, and thus poor and unsatisfactory fluid arises.

To stamp out tuberculosis, three methods have been advocated:

1. The slaughter of all cattle reacting to tuberculin (an impracticable measure). Even the separation of the healthy from the diseased ones, as indicated by the tuberculin test, would at present be a tall order.

2. Scheduling udder and open tuberculosis, slaughter, and compensation — a reasonable and valuable suggestion which could be carried out.

3. The rearing from birth of cattle free from the disease, and giving them every opportunity of living lives as far as possible free from infection. Practically no calves are born with the disease; they all acquire it. Keep them free from infected stock, and tuberculosis will not develop among them.

The tuberculin test, when applied to individual herds, is a measure highly to be commended, but the time for it to be universally adopted has not yet arrived. The writer has seen bulls which, on *post-mortem*, have been found badly affected with tuberculosis. They have been procreating their species for some time while in this condition, and have probably helped to infect the dams.

The tuberculin test, at any rate, might be applied to many of them with advantage.

Mr. John Burns' recently introduced Milk Bill and the Tuberculosis Order of the Board of Agriculture provide for the notification, slaughter of, and compensation for animals suffering from visible or advanced tuberculosis. The coughing, emaciated cow with hard lymphatic glands or udder ought, therefore, soon to be extinct in the cow-sheds of our land. The order says the milk from such a subject should be boiled, nothing further. Presumably, after boiling, the fluid may be given to the pigs.

DISEASES OF COWS THAT MAY AFFECT MILK

Mammitis, or Garget

Inflammation of the udder is a very frequent complaint of milk cattle. The liability to it cannot be wondered at, considering the periodical changes the gland often undergoes—at one time full of blood and flushed with milk, at another quiescent and dry. Two chief kinds of mammitis occur: the one affecting the lining membrane of the udder, and commonly producing a catarrh; and the other attacking the substance of the gland itself, and leading to pus formation and sloughing or death of the quarter attacked. It only requires exposure to cold air or water or wet shippon floors, when the bag is heated or in full play after calving, to bring about its inflammation. This generally leads to the catarrhal form of the complaint. Blows, injuries, infected cracks or sores on the teat, frequently occasion the deep-seated form of the disease, and an abscess not uncommonly arises. Udders may be trodden on by one member of the cow-house stamping on the gland of a recumbent companion, or a cow with a very pendulous bag may injure it herself. A severe attack of garget may arise thus.

Milk imperfectly removed from the udder becomes subject to decomposition; bacteria grow in it, and catarrh is set up. The little drop of milk left hanging on the teat of a cow after milking, or the fluid dried on where the teat has been wet with the secretion, may favour the entrance of bacteria up the milk-duct. Some organisms probably have a power of forcing their way through the little passage in the teat up into the vessel, and there, coming in contact with some milk left behind, cause it to decompose, and set up mischief. In this connection it will be seen how important it is that milk catheters should be boiled and absolutely clean before insertion. Milk left in a healthy gland never of itself sets up inflammation, but it is a splendid seed-bed for organisms once they gain access to it, and they frequently do reach it, especially in hot weather. A stiff piece of dirty straw penetrating the opening in the teat when the animal is lying down may help the bacteria to gain entrance to the bag.

The udder symptoms of catarrhal mammitis where the lining membrane is attacked are that swelling and redness may be slight, but there is pain on pressure, and irritability during milking. Pus or matter may occasionally form, and the tissue of the gland be attacked.

In the case where the substance of the bag is affected, all the symptoms are more severe. There is great pain, swelling of the gland, heat, and redness of the skin; the cow loses her appetite, there is fever, thirst, and depression; an abscess forms, and sloughing or death of the quarter or quarters attacked may occur.

A cow that has been the subject of garget frequently loses the milk from one quarter of her udder; thus it is economically important to do everything to prevent this disease.

The Effects of Mammitis on the Milk are that the secretion may contain blood, be flaky or clotted, thick, slimy or purulent, from the action of the microbes and their products. Such milk should not be used, as it may be especially dangerous to young children, causing inflammation of the stomach and intestines.

To prevent mammitis occurring, the cow-shed floors should be kept clean and dry, and the diseased gland never milked on the ground. The animal should not stand in a draught. Cows with any discharge from the shape should be excluded from the shippon. The milker, in order to avoid carrying infection, should be very particular to wash his hands after milking an affected cow, a 3 per cent. carbolic acid solution or creolinized water being used.

There are several germs found in the milk of cows suffering from mammitis. The chief ones are those in chains (streptococci), and clusters or bunches (staphylococci). These organisms, when taken by human beings in milk, may cause fever, weakness, prostration, vomiting, diarrhœa, and crampy pains in the limbs. Outbreaks of sore throat have also been attributed to such milk, similar bacteria having been found in the pharynx of human beings drinking the fluid.

Cow-pox

Cow-pox is due to a virus or organism, which attacks the body of the udder and the teats. There is often slight feverishness in the animal attacked, but these symptoms may be so mild as to pass unperceived. The teats or parts of the udder become swollen and warm; lumps or knots form on them about the size of a pea, or smaller. The lumps soon fill with fluid called lymph, and little blebs, blisters, or vesicles arise, which are round or oval. These ripen, sink in the middle, burst, dry, and a scab forms, which is called a pock. The number of these pocks may be one, two, or fifteen to twenty. To the sight, and when unabraded, they are bluish-white, with a metallic glaze and a congested area or ring round them, called an aureola. Any scab forming over the end of the teat and blocking the canal will retain the milk and be troublesome. It should be removed. If the cow's bag has been handled much, the pocks do not always assume the regular and classic form seen in the undisturbed eruptions. They may be raw erosions, or irregular-shaped sores cracked across, and with a profusion of easily removed, softish, brown scab.

Preventive Measures.—The floor and litter where the cow stands should be kept as dry as possible, the udder and teats likewise. Dry milking is better than wet if it is desired to give the complaint least chance of occurring. When the disease has arisen, the bag should be handled cautiously. The milk-siphon may be used to draw off the secretion, as by its use the teat may be kept fairly still, and the least chafing of sore places occasioned. All milk-tubes should be inserted carefully, and well cleansed and boiled before using. It is very important that cows with any sores on their udders should be milked last in the herd, and the cowman should wash his hands in disinfectant after handling a diseased udder. Thus milkers will avoid carrying the disease from cow to cow. Infection may be carried through the bedding, and it should be looked to, and the cows kept up from pasture. Outbreaks of cow-pox often occur where the animals are too closely confined in dirty sheds, and where milkers are not very clean in their habits, and, above all, does the disease quickly spread when the ailing cows are milked between the healthy ones.

It will easily be understood that the milk from an udder affected with cow-pox should not be used. It may contain the discharge from the bleb or sore, or bits of scab from the pock, which it is very undesirable that human beings should drink. As this disease usually lasts about three weeks, the milk-supply should be discontinued for that time, and until the pocks are quite healed. Animals affected should be isolated, and the fluid from them may be boiled and given to pigs or poultry. Children that have partaken of cow's milk containing the virus suffered from eruptions and scabs on the face.

Foot-and-Mouth Disease

This complaint nowadays only occasionally visits these isles. It is shown by blebs, blisters, or vesicles which affect the mouth, the space between the claws, and the udder of milch cattle, causing sores, inflammation, or abscesses of the vessel. The cows smack their lips and dribble at the mouth. The vesicles spread, and may involve the coronet, or in acute cases the whole of the digestive tract from the anterior opening, the mouth, to the posterior one, the rectum.

The blebs running into large sores may cause much pain or death. In severe cases the milk of dairy cows becomes thin, bluish, and poor in fat. Where the udder is badly affected the secretion may be yellow, dirty, and blood-tinged. On standing, a layer of cream quickly forms, but the milk presents a bad-smelling, slimy mass, having a nauseous rancid taste. This disease may be transmitted to man by raw or uncooked milk or its products. In rare cases men have been infected by handling animals subject to the disease. The complaint runs a mild course in adults. In children catarrh of the stomach and death may occur from consuming the milk.

Several cases have been reported in Germany of children showing blisters on the tongue and lips, running into sores. The virus of foot-and-mouth disease is easily destroyed by heat of 100° C. Milk thus boiled loses its nourishing power, however, so that there should be no consumption of the fluid in this complaint.

Actinomycosis

"Wooden tongue" is the common name for this disease. The complaint, due to an organism called the ray fungus, found on barley-awns, affects the tongue, jaws, or throat of cattle, and occasionally the udder of milch cows. It is shown by dribbling of saliva, pain in feeding, sores on the tongue, especially at its edges, and swellings on the jaws or in the throat. When it affects the udder of a cow it is denoted by little hard lumps in the bag, which may discharge pus. These nodules generally occur most about the base of the teats.

Another affection of the udder, resembling this one and but rarely seen, is called *botryomycosis*, where there is great thickening of the udder and burrowing passages or sinuses running into the substance of the gland, from the openings of which pus or matter escapes. Both these diseases of the udder resemble tuberculosis somewhat in outward appearances, and only the microscope and the skilled veterinary surgeon can determine which is which.

No cases of communication of these diseases to man through milk have been recorded, but it is possible that the human subject might be affected through the digestive tract, and, in any case, the milk should not be used, as it frequently contains pus.

Anthrax

This serious blood disease, due to a rod-shaped organism, called the *Bacillus anthracis*, soon destroys life. A cow affected shows almost immediate cessation of milk secretion. The fluid becomes a yellowish colour, or intermingled with blood. The albumin is diminished, and sugar and fat increased. The illness is transmissible to man through the organism, and anthrax bacilli may be present in the milk, especially when bleedings take place in the udder. In this way calves may be infected.

In man, infected through his digestive tract, there will be violent bleeding, inflammation of the bowels, ushered in with great stomach pain, vomiting, and shivering; finally, collapse and suffocation. The organisms may live in drawn milk for fifteen days. Only few cases have been recorded where men have taken the disease by consuming anthraxed milk, probably due to the fact that milk secretion often suddenly ceases, and the disease gets to the udder in the last stages of the attack, when all precautions have been taken and the illness notified under the Anthrax Order.

Rabies (Hydrophobia)

In a cow bitten by a mad dog, and thus inoculated, the virus of this disease might be secreted in the milk. Milk from an affected source inoculated into experimental animals produces the disease, but a similar supply given to calves has been resultless. A case has been recorded where a woman affected with hydrophobia suckled her child, and no harm occurred to the infant. It is quite possible, however, that through little sores, ulcers, or fissures in the mouth, stomach, or bowels, the virus might enter the system and set up the disease. The sale or use of milk from a rabid animal is rigidly forbidden.

Illnesses of the Digestive Organs

The cow suffers from many forms of digestive trouble. In tympanites or hoven, when an animal gets blown up like a balloon, gases are absorbed from the stomach, and the milk may smell badly. In liver disease, impaction of the third stomach, and constipation, the fluid will become tainted. In diarrhœa or gastro-enteritis bacteria or their products (toxins) gain access to the blood-stream, and so come into the secretion.

The coli group of organisms, always found in these diseases (present also in ptomaine poisoning by sausages), may get into the pail during the milking act, especially if the tail of the animal gets soiled by loose, semi-fluid, or blood-tinged fæces constantly running away from the cow at milking-time.

Milk from animals thus affected, when consumed by man, may cause faintness, headache, and shivering, followed by diarrhœa, vomiting, and high fever. Much illness among calves is caused by the bacteria found in cows suffering from gastro-enteritis, and not only should such milk be kept from the youngsters, but the sick should be put right away by themselves, and the attendant, if he also feeds the calves, should be clean and careful in his methods.

In septic illnesses of the womb or vagina after calving, and in contagious vaginitis of the cow, the milk may become contaminated. The foul secretions coming from the opening of the vagina soil the parts and tail, and are thence conveyed to the udder and under surface of the body, and so into the milk-pail. In diseases of the bladder, and where a foul-smelling cleansing or afterbirth is retained, gases may form which injuriously affect

DISEASES OF COWS THAT MAY AFFECT MILK

the fluid. In all wounds of the cow's udder or contusions, and in injuries on the body where pus is formed, there is a danger of stuff getting into the milk and causing serious illness in children.

Milk Fever

Toxins are formed in the udder in this complaint, and early milk drawn from a cow suffering from the illness and injected into the jugular vein of another animal caused death. Experimental animals, such as rabbits, also died, showing stiffness and cramp. The milk on standing has a big layer of fat, and is watery and bluish-white underneath. Through long lying, tympanites in the subject may accompany this complaint and injuriously affect the milk.

Red-water

This is a blood disease occurring in certain neighbourhoods, especially on coarse, swampy, and forest pastures. It is well known in the West of England and in some parts of Ireland. The complaint is prevalent as a rule in spring and summer. The organism that causes the illness is ring or pear shaped in form, and is parasitic on the red blood-corpuscles, which it breaks up. The temperature of an affected animal rises, and the urine becomes red in colour, and contains the colouring matter of the red blood-corpuscles—viz., hæmoglobin.

It is rather interesting to know the way infection is believed to be spread in this complaint. It is by the agency of blood-sucking creatures called ticks, which are usually to be found under the forearm, inside the hind-leg, or on the udder of a diseased animal. These ticks live on the vital fluid of the subject, and, having absorbed their fill from the blood containing the organism, they inoculate a fresh subject during the blood-sucking process, either directly or by means of the infected developing offspring.

There is a form of red-water intimately associated with parturition in poor, badly nourished cows. It is hard to say definitely that ticks propagate the disease here; but the urine is often as red-coloured as port wine, the animal's temperature high, and the milk discoloured and watery.

The secretion of lacteal fluid diminishes rapidly in red-water. The supply is yellow in colour, bitter in taste, and unfit for human consumption.

Abortion

Slipping the calf is due to an organism that gains access to the genitals directly through the external opening or vulva, or frequently by being swallowed with the food, or taken in by the tongue, being absorbed through the intestine, and passing into the blood-stream.

It usually occurs in cows from the fifth to sixth month of pregnancy, and is indicated by a white, reddish-grey, or yellow discharge from the vulva. It may be spread from cow to cow by contact of the secretion, or the bull may transmit it from a diseased cow to subsequent animals he serves. The flow of milk is seriously interfered with, and the bad discharges arising from the genitals render special care necessary in dealing with such animals. Where they cannot be isolated, if possible, they should be separated from the rest of the herd in the shed, and put by themselves at the lowest place in the byre, where the discharges from their hind parts will run right away, and not come in contact with those of other pregnant cows. Isolation should last for three months, and the contaminated parts of the body be washed with a disinfectant, so long as the bad flow soils them. The feeding-trough and litter should receive special attention in this disease, and everything be done to prevent contamination of the food and water supply by discharges containing the bacilli.

Germicide vaginal injections and cleaning of the back parts are copiously indicated, and all the measures recommended under "Checking the Spread of Disease" should be practised. Fœtus, cleansings, and bedding should be burnt. Destroy the first if born alive, and burn or bury deeply in lime. The milk in abortion does not vary much in composition, but supply is usually seriously diminished, and present calf-bearing value reduced to nil. Inasmuch as the abortion bacillus may easily get into the milk of a cow suffering from this complaint, it would seem a safe plan to discard and destroy the milk, and not run the risk of conveying the disease to other pregnant subjects by it. In any case, exclusion of the fluid from human consumption is desirable. It is hoped that a means of conferring immunity from this disease may be established by the injection of healthy cows with cultures of the abortion bacillus. Gratifying experiments in this direction have been conducted at the Board of Agriculture station at Alperton by Mr. Stewart Stockman, M.R.C.V.S., and the Abortion Committee. The cows are inoculated, or injected subcutaneously.

CHAPTER XXI

CHECKING THE SPREAD OF DISEASE

There are three chief matters to be considered in any attempt to stop the advance of disease. In the cow-shed the sick animal is a centre and source of the complaint; her surroundings, and possibly the air around, are infected by her, and any live or dead thing may carry and transmit the germs of the malady from her to another creature. The animal with open tuberculosis is a focus of disease; she soon infects her surroundings; whilst the virus of cow-pox is often conveyed from cow to cow, and the organism of mammitis is spread by litter.

All effective fight against disease is really a matter of killing bacteria, and wherever these minute bodies exist, grow, or spread, they must be attacked and slaughtered. If a battle is to be brought to a quick ending, an astute leader pens the foe within a limited area. Having confined the enemy, the fight can be carried on in a much smaller compass; and if the attacking force be stronger, capitulation will be in sight. It may be taken as a sound axiom that to isolate or put by herself any diseased cow, or, indeed, any ailing animal, is the first sure step to be taken in checking the spread of disease, as it is by so doing that the greater majority of the bacterial foe is confined. The patient should not be put into a dark, dirty, or damp place, but in as clean and airy a one as possible. Thus the originating germs will be put in one house—under cover, as it were—and their gambollings can be limited; whilst the standing the patient has left, where evil organisms left behind dwell and lurk, can be attacked with disinfectant, which is the arch-enemy of all bacterial life.

On not a few homesteads it is difficult to isolate cows suffering from illness; but cleanliness should be rigidly observed in and about them, and some sort of contrivance (such as a sheet or sacking soaked in a germicide) rigged up to partition them off from their companions, and disinfectant used. A good inodorous germicide used in French cow-houses is called "eau de Javel." It consists of 3 ounces of chlorinated lime dissolved in 4 pints of water; then add 3 ounces of pearl-ash, filter, and put 2 drachms of hydrochloric acid in the mixture, and shake. Having put the patient by herself, the stall or standing she has occupied should be washed out with a bacteria killer, and manger, stall-posts, walls, and anything the animal has come in contact with, or her dejecta reached, scrubbed with hot water and disinfectant, and re-limewashed. Before using a disinfectant, brushing surroundings with boiling water charged with carbonate of soda is desirable. An empty stall between the one recently vacated and the next occupant of the shed is good practice. A 20 per cent. carbolic solution, or 2 lb. of sulphate of copper in 3 gallons of water for woodwork and floors, or a strong solution of a coal-tar disinfectant, are suitable germicides. Powdered lime or limewash sprinkled on the floors keeps them sweet and clean. The roof and top portion of the walls may be sprayed with a germ destroyer. The

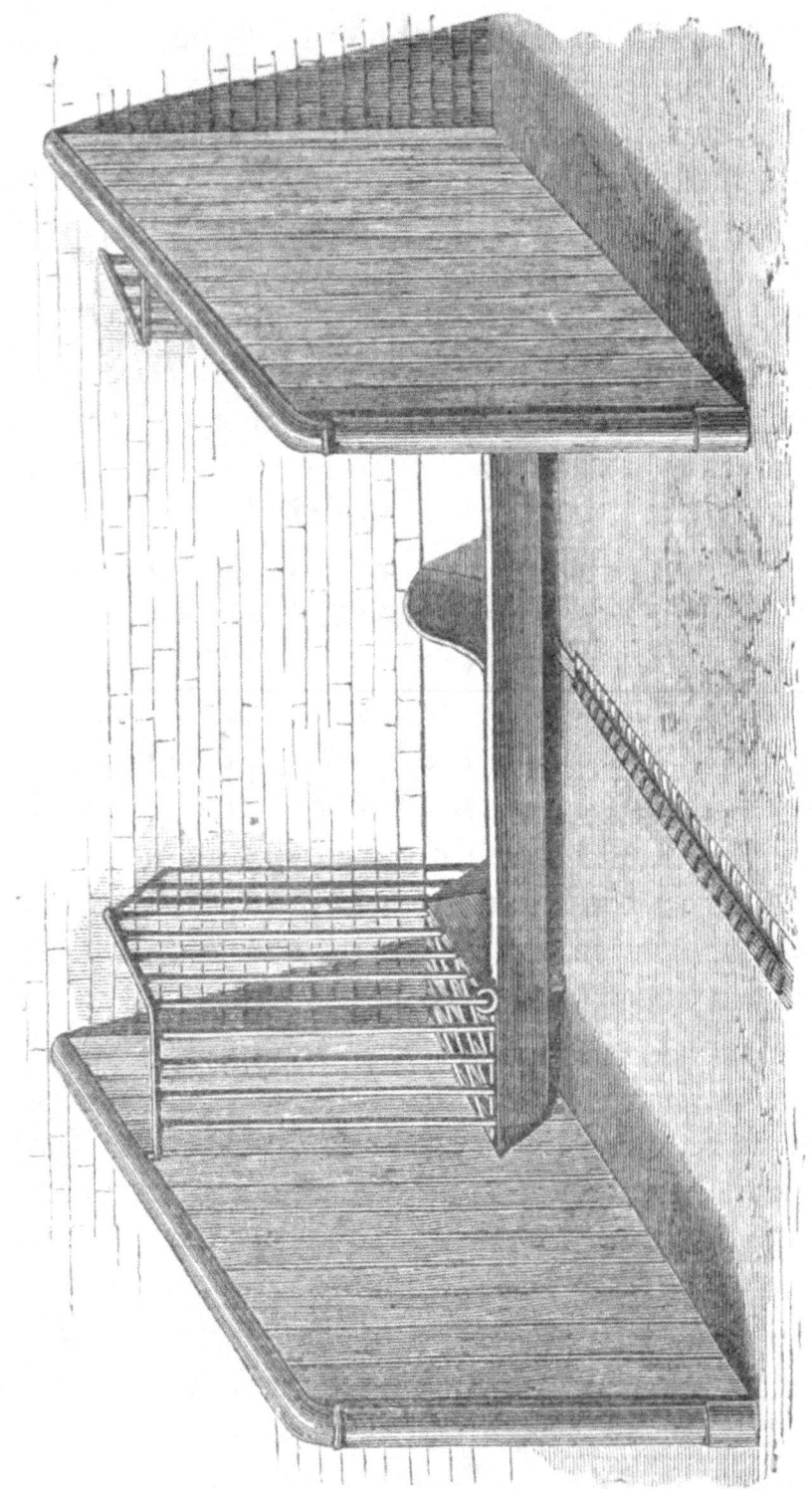

Fig. 7.—Fittings with Divisions of Wood, and Iron Posts and Rails. (St. Pancras Ironwork Co., Ltd.)

litter should be burnt. Infected dung needs safe disposal, and should not be used as manure, and contagious genital discharges require disinfection and washing away. Having cleansed and disinfected the stall and things near it, it should be left empty for a time, and, if possible, fresh air and sunlight plentifully admitted to it.

It is often forgotten what a destroyer of germs the glare of the sun is.

In a recent experiment in India, where the sun tem-

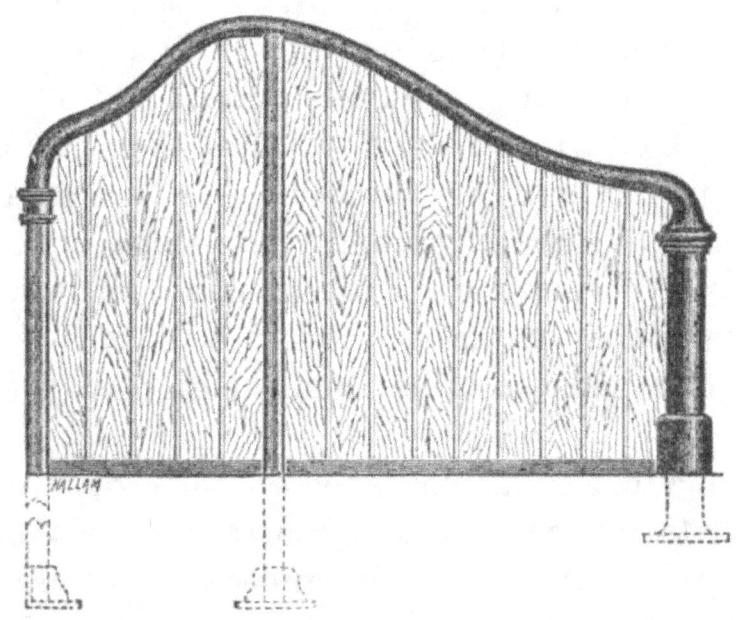

Fig. 8.—Wood Stall Division: Iron Posts and Rail.
(St. Pancras Ironwork Co., Ltd.)

perature was 105° F. and the shade reading 92° F., the typhoid bacillus was killed by an exposure of one hour to the sun; 240,000 bacilli were killed entirely after being subject to the sun's rays for three hours and a half. In clear water sunlight will kill germs to a depth of 6 feet (R. T. Hewlett, M.D., Milroy Lectures).

Having seen to the standing, and isolated the sick animal, the prevention of the carrying or transmitting of disease germs by persons, animals, or things should occupy attention. The infection may be carried from cow to cow, and, in some cases, from the ailing creature to himself (cow-pox) by the attendant. Such a one, if obliged to attend healthy stock after seeing to the wants of the patient, should be clean in his habits and person. He should wash his hands in antiseptic, brush his clothes, in some cases disinfect his boots, and go out into the pure breezes before walking in and out among the healthy milking herd. He should let as long a time as possible elapse between caring for the sick and the period he visits the cow-house. Dogs and foxes may spread the contagion of abortion. Separate buckets and pails should be used for the ailing cow, her own brush if grooming is carried on, and one sponge or cloth for her udder. Whenever bad discharges soil the hind parts of a cows these should be washed with a 2 per cent. solution of creolin, or 3 per cent. of carbolic acid twice daily.

All flat and smooth things with impervious surfaces are better to conduct the slaughter of bacteria on than rough or permeable stuff, with chinks, corners, or crevices in it. Hence impervious floors are better than those of soil; smooth plaster or polished brick surfaces than those of rubble or rough stone; ironwork and earthenware mangers than old wooden cribs and posts.

When the patient gets well, the isolated loose-box or stall she has occupied should be dealt with by cleansing, disinfecting, whitewashing, and leaving empty to admit sunlight and fresh air. The walls may be limewashed and floor disinfected periodically while the patient is in the box, if it is necessary to isolate her for a long time. It is always wiser to keep a cow by herself over her time of recovery than to be in a hurry to put her back among companions.

The conferring of immunity from disease by the use of vaccines, sera, and the products of bacilli, has perhaps received more attention in foreign countries than here, as a method of checking the spread of contagion, but there is hope that the losses from abortion may be reduced in the future by preventive inoculation.

LITERATURE

Variation of Animals and Plants under Domestication. Charles Darwin.
The Farmer and Stockbreeder.
The Field.
The Journal of the Board of Agriculture.
The Veterinary Journal.
The Veterinary News.
The Veterinary Record.
Walley's Meat Inspection. Edited by Stockman.
Cattle: Breeds and Management. By Housman.
British Flora. Bentham and Hooker.
Veterinary Obstetrics. By George Fleming, C.B., LL.D.
First Lessons in Dairying. Hubert van Norman.
Milk: its Nature and Composition, C. M. Aikman, M.A., D.Sc.
Modern Dairy Farming. Puxley.
Handbuch der Milchkunde. Rievel.
Spezielle Pathologie der Haustiere. Hutyra and Marek.
L'Hygiène de la Viande et du Lait.
Revue Générale de Médecine Vétérinaire.

INDEX

Abnormal milk, 75
Abortion, 92
Actinomycosis, 88
After-birth, removal of, 34
Age of cows, 28
Air-space, 43
Anthrax, 88
Average yield of a good dairy cow, 24
Ayrshires, 8

Backhaus's children's milk, 61
Bacteria and milk, 63
Bad-smelling milk, 75
Bad-tasting milk, 76
Bitter milk, 76
Blue milk, 75
Boracic acid in milk, 67
Breeds of milk cattle, 5
Buddeization, 67
Bull, 14
Butyrometer, 58

Calves and their rearing, 36
Cattle breeds and their evolution, 1
Clean milk, 69
Cow, a fine-quality, 13
Cow-house, improving the old, 47
Cow-houses or byres, 40
Cow in milk, 23
Cow-pox, 85

Damp and dry milking, 69
Defects of teat canal, 54, 55
Devons, 9
Dexter Kerries, 8

Disease, checking spread of, 94
Diseases of cow in which milk should not be used, 70
Drying the cow, 20
Dutch cattle, 9

Effects of mammitis on milk, 84
Evolution of cattle breeds, 1

Feeding for milk, 17
Floors of cow-houses, 42
Food for children's cows, 62
Foot-and-mouth disease, 87
Formalin in milk, 67

Grooming the cow, 69
Guernseys, 7

Hereford cows, 9

Illnesses of the digestive organs, 90
Improving the old cow-house, 47

Jerseys, 7

Kerries, 8

Lactometer, 58
Lighting, 43
Literature, 100
Litter, 73

Making milk, 17
Mammitis, 83
Milch cattle, 11
Milk, 56

Milk fever, 91
Milk for children, 61
Milk-house, 74
Milk secretion, 12, 54

Pasteurization, 67
Preservation of milk, 66
Preventive measures in cow-pox, 86

Rabies, 89
Red milk, 75
Red polls, 6
Redwater, 91
Removal of after-birth, 34
Reproduction, 31

Salicylic acid in milk, 67
Septic illnesses affecting milk, 90

Shorthorns, 5
Slimy milk, 75
Stalls, 44
Sterilization, 67
Structure of the udder, 54

Tuberculosis, 77
Tuberculosis of the udder, 80

Udder tuberculosis, 80
Utensils, 72

Variations in milk, 25
Ventilation, 43

Wash-house, 74
Water-supply, 71
Watery milk, 76
Welsh cattle, 8

THE END

BAILLIÈRE, TINDALL & COX, 8, HENRIETTA STREET, COVENT GARDEN

Plate XV.

Photo. G. H. Parsons.

WELSH HEIFER, MADRY SALLY.
First, Royal Show, 1907 and 1908.

Plate XVI.

Photo. G. H Parsons.

RED POLL COW, DESIRÉE OF JOHNSTOWN.

Plate XIII.

Photo. G. H. Parsons.

JERSEY COW, LADY VIOLA.

First, Royal Show, 1906, 1907, and 1908.

Plate XIV.

Photo. G. H. Parsons.

GUERNSEY COW, ICHEN DAIRYMAID.

First, Royal Show, 1908.

Plate XI.

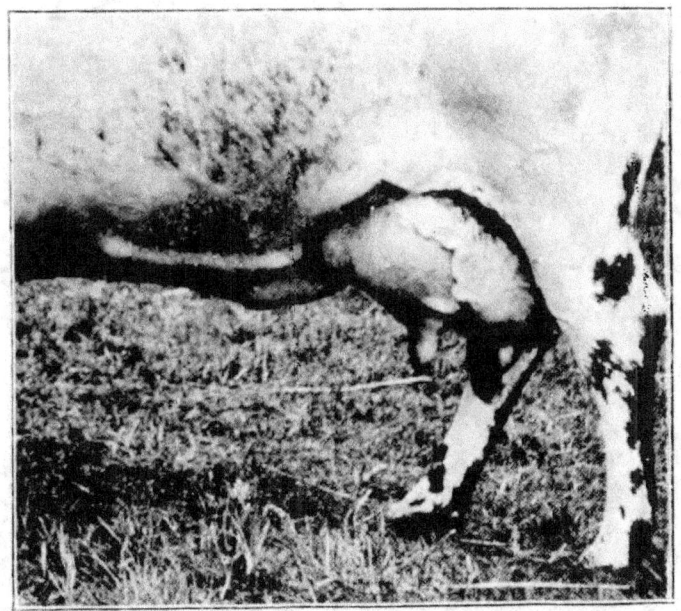

Photo. G. H. Parsons.

A GOOD UDDER.

Plate XII.

Photo. G. H. Parsons.

DEXTER COW, LA MANCHA, HARD TO FIND.

Owned by Hon. Mrs. C. Portman, Stratford-on-Avon.
First, Bath and West, 1908; First and Champion, Royal Counties, 1908.

Plate IX.

Photo. G. H. Parsons.

LINCOLNSHIRE RED SHORTHORN COW, KEDDINGTON SKIPWORTH, AND CALF.
First and Champion, Royal Show, 1907.

Plate X.

Photo. G. H. Parsons.

Plate VII.

Photo. G. H. Parsons.

FULL PEDIGREE DAIRY SHORTHORN COW, GIFT II.

First, Royal Show, 1908.

Plate VIII.

Photo. G. H. Parsons.

NON-PEDIGREE DAIRY SHORTHORN COW, MINNIE.

First, Dairy Show, 1908.

Plate VI.

Photo. G. H. Parsons.

SHORTHORN BULL, METEOR.

Owned by Sir Richard P. Cooper, Bart., Shenstone Court, Lichfield.

To face page 16.

Plate IV.

Photo. G. H. Parsons.

DEVON COW, WHIMPLE KITTY.
First, Royal Show, 1908.

Plate V.

Photo. G. H. Parsons.

DAIRY SHORTHORN BULL, WESTON PRIDE.
Owned by Mr. Thomas Parton, Weston Hall, Crewe.

Plate II.

DEXTER COW, COMPTON DOLLY VARDEN.

Owned by the King.   First and Champion, Royal Show, 1908.

Plate III.

KERRY COW, BELVEDERE NORA.

Owned by Mr. J. L. Tillotson.   First, Royal Show, 1908.